British Agricultural Cooperatives

JOHN MORLEY

British Agricultural Cooperatives

HUTCHINSON BENHAM, LONDON

HUTCHINSON BENHAM LTD
3 Fitzroy Square, London W1

An imprint of the Hutchinson Group

London Melbourne Sydney Auckland
Wellington Johannesburg Cape Town
and agencies throughout the world

First published 1975

Set in Monotype Imprint

Printed in Great Britain by The Anchor Press Ltd
and bound by Wm Brendon & Son Ltd
both of Tiptree, Essex

ISBN 0 09 123960 5

Contents

Foreword

by the Chairman of the Central Council for Agricultural and Horticultural Co-operation

This is the first general book on its subject for many years. It was time for one to be written. Agricultural cooperatives have changed greatly, even during the seven years I have known them. I have no doubt that they must and will change even more during the next decade, when they will be faced by many new challenges.

They will also have exceptional opportunities. How else, except through cooperation, can the agricultural industry keep its character and independence? In this troubled second half of the twentieth century it is only by working together that mankind will be able to survive.

If they are to fulfil the high hopes entertained for them, agricultural cooperatives must possess leaders gifted with imagination, and backed by a sound understanding of the possibilities inherent in cooperation as a means of development. Where such inspired leadership exists, there is seldom any real difficulty in finding material resources to support it. Since its formation in 1967, the Central Council has sought to take every opportunity it could find of educating and training a new generation of members and managers. It has given this task equal priority with the administration of the grants scheme and its promotional work in general.

This book is basically about the existing situation of British agricultural cooperatives, where they stand at this point in time, and how they arrived there. But it will not be difficult, I think, for the reader to draw some conclusions about the way that they should progress in the future, and it is, of course, the aim of the book that he should do so. It was not commissioned by the Council but is – to use a well-worn, but unavoidable expression – a labour of love. The Council has gladly sponsored its publications in the belief that nothing but good can come from wide discussion of the points of view, many of them controversial, which are expressed in its pages.

John Morley, the author, has been a welcome colleague for many years. No one is better qualified to have put together the material that is now presented.

March 1975 ROGER FALK

1. Economic and political background

The agricultural cooperative situation in a country is a compound of its past history and its present opportunities. The history of agricultural cooperation in the United Kingdom will be the subject of the next chapter, while its opportunities are the theme of the book as a whole. To set the stage for this discussion, it will be useful to begin by drawing attention to several features of the contemporary agricultural economic scene in the United Kingdom and, so far as they may be relevant, in the area of the European Communities generally, where cooperation appears to have a special significance. The object of doing so is to enable the reader, at the very start of the present study, to come to some preliminary conclusions about the theoretical advantages of cooperation among farmers in present-day conditions, which will be of value to him when looking later at situations in which cooperatives are actually involved.

Information is published annually by the Ministry of Agriculture, Fisheries and Food on the *structure of agriculture in the United Kingdom*, where the number of holdings considered to be statistically significant in 1973 was 280000. In a review of the changes which had taken place between 1963 and 1968, an earlier report had referred to a trend towards concentration of enterprises, and this has clearly continued into the seventies. Enterprises are becoming concentrated in two ways; there are fewer holdings, and a minority of them account for a major part of domestic production. Thus, in 1973, under one-third of the total number of holdings, with a labour requirement of 600 or more 'standard man days', accounted for more than three-quarters of the total output. It is evident that, in such circumstances, there may be the alternative objectives open to an agricultural cooperative, of attracting as many producers as possible into the organization, or of obtaining as large a share as possible of the available production.

Needless to say, the size of holdings is far from uniform in different parts of the country. There is a smaller than average size of holding in Wales, and it is smaller still in Northern Ireland. Likewise in certain crops, such as cereals, potatoes and sugar beet, the proportion of the national output concentrated in the larger holdings is relatively high, while in the case of livestock enterprises the distribution of production among holdings is more even. In practice, therefore, there are mitigating factors which will help to resolve a possible conflict of cooperative objectives, even though they may not entirely remove it.

The *size of holdings in other EEC countries* is generally much smaller than in the UK; this fact has been cited as one of the many reasons why agricultural cooperation developed differently there. In the United Kingdom, according to the European Communities Commission's annual report of 1973 on the situation of agriculture, which for the first time included detailed statistical information about the enlarged Community of nine members, as well as about the original 'Six', 27·3 per cent of UK holdings (amounting to 80 per cent of the farmed area) were rated as being in the category of fifty hectares (120 acres) or above, compared with 5·2 per cent in the 'Nine' (the figures being for 1970 in both cases). One in four of all such holdings was located in the United Kingdom.

From the same source evidence can be taken of *proportions of agricultural to total population.* This too may be relevant to the situation of agricultural cooperatives since it would seem to be logical for farmers to seek to preserve through better political and economic organization the strength they are in danger of losing through decline in their voting power. The figures are revealing; during the four years 1969–72 the part of the population employed in agriculture, forestry and fishing in the 'Six' fell from 13·8 to 11·2 per cent; the corresponding figure for the 'Nine' in 1972 being 9·4 per cent and for the UK 3·3 per cent. Looked at over a ten-year period the change is even more striking, as appears from a comparison of the number of the agricultural population in the 'Six' in 1970 with the corresponding number for 1960; the figure is reduced to about half what it had been formerly.

A study of the proportion of the *gross national product attributable to agriculture* might lead to the same conclusion, that the industry needed to take steps to organize its own protection. Here the figures for the 'Six' show a decline between 1969 and 1972 from 6·6 to 5·7 per cent; in the fourth of these years the corresponding percentage for the 'Nine' was 5·3 per cent and for the UK was 2·8 per cent.

Again, any signs of a deterioration in the *terms of trade* as between agriculture and other sectors of the economy must cause misgivings, and encourage farmers to look for remedies in the form of better economic organization, including cooperation. The Commission's 1972 report said in this connection: 'The development of the relationship between prices received and paid by farmers, otherwise known as the terms of trade for agricultural products, constitutes an important element in determining the development of agricultural incomes. By reason of the lack of appropriate statistics, these terms cannot be worked out so as to take account of the costs of all means of production . . . We may note however that in each of the member States the terms of trade for agricultural products are inferior to those of the year taken as base (1966 or 1966–7).'

Consideration of the terms of trade involves the relationship between farmers and their suppliers on the one hand, farmers and their customers on the other. The question of *farmers' 'buying power' in respect of inputs* is one that has played an important part in agricultural discussion of recent years. The main inputs are feedingstuffs, fertilizers, and machinery (repairs, fuel and oil, and other non-capital costs) which, according to the United Kingdom *Annual Review of Agriculture 1974*, accounted for approximately 70 per cent of farm expenditure in 1972–3. So far as feedingstuffs are concerned, although the number of mills with over 10000 ton output increased slightly in 1970–1 (the latest year for which figures are available), due to the continuing switch of milling capacity from deep-water ports to smaller ports and inland areas, nearer to cereal supplies and livestock production, at the same time the control of these manufacturing sources became more concentrated. According to estimates made by Professor Britton in his study of *Cereals in the United Kingdom* (1969), some 60 per cent of the production of compound feedingstuffs was handled by eight firms; currently, the proportion of trade controlled by the 'nationals' is thought to be nearer 70 per cent. As regards fertilizers, the *Agricultural Merchant* of August 1969 calculated that three firms were responsible for 75 per cent of the total business. This might now be nearer 85 per cent if straight nitrogen as well as compounds were included. As regards machinery, much of the agricultural repair bill consists of spare parts of machinery coming from a fairly small range of manufacturers, while the concentration of firms in the fuel and oil sector is well known. When one relates these figures of sources of supply with those of the number of farm holdings quoted earlier, it is not surprising to find that farmers have come increasingly to doubt their ability to negotiate favourable terms on an individual basis.

The situation of farmers in *relation to their markets* is considerably more complex than that described above. In time to come it will be necessary to consider this situation in its European terms, but for the present it makes better sense to do so on a more local basis. Moreover, facts concerning food-marketing trends in the United Kingdom are conveniently available from a paper on this subject which was given by Dr P. A. Power of the Ministry of Agriculture, Fisheries and Food to a CENECA (Paris) International Symposium in 1972. What follows is taken from this paper. Consumers' expenditure on food declined slowly but steadily during the past decade; from being almost 29 per cent of total expenditure in 1960 it had fallen to 23½ per cent by 1970. Such a trend is, of course, a familiar phenomenon in countries where there is a rising disposable income per head of the population. Difficulties arise in evaluating it, however, because the United Kingdom food-supply industry is far from being synonymous with United Kingdom producers. It comprises not only these producers, but also the importers who make available non-indigenous and imported food supplies, the food manufacturers who prepare and process the raw food, and a large number of distributors who undertake the transporting, storage, wholesaling and retailing functions which comprise the marketing process. Two conclusions emerging from this survey are that during the period 1962–3 to 1969–70 home agriculture marginally increased its share of both the value of domestic food supplies and the total consumer expenditure on food, and secondly that, throughout the same period, agricultural producers and importers between them on the one hand, and food processors and distributors on the other, each accounted for about half of the annual value of the market for food in the United Kingdom. The first of these conclusions is a reminder of the fact that before entry of the UK into the European Communities (but the position has not basically been changed by entry) the domestic food market was shared between supplies from home and abroad, to the extent of making it difficult for the home-based industry, however well organized, to exercise a dominating influence. The second seems to show that the procedures for marketing, if the costs of it remained so remarkably constant over a ten-year period, must have been reasonably efficient; though it must be added that it was not possible to distinguish the costs incurred in marketing home-produced food from those incurred in marketing imported food products, and one of these operations may well have been more efficient than the other.

The question of *vertical integration* in agriculture has been much considered in recent years, mostly in relation to marketing, though this relationship may also be found between agriculture and the industries that supply it. Agricultural producers, as pointed out in the CENECA paper mentioned earlier, are largely concerned with the pre-marketing functions of growing food, but they also undertake in certain instances some of the processing and distributive functions. Likewise it is not unknown for processors and distributors to involve themselves in pre-marketing and even production activities. The balance between production and marketing activities has been changed in the recent past, as a consequence of what is described in the CENECA paper as 'the growing tendency on the part of the housewife to buy her own time by transferring to the factory much of the work that she used to do in the kitchen' and to demand 'improved presentation of the food in terms of improved packaging and more attractive display'. During the 1964–9 period household expenditure on convenience foods increased at an annual rate of almost 6 per cent compared with a growth rate of just under 3 per cent for other foods. The burden of this added cost falls mainly on the processor. These trends, the CENECA paper suggested, are likely to continue in future years, when the increasing demand for inbuilt services, more preparation, more expensive packaging and presentation of food products will continue perhaps at a more rapid rate than in the past, and the value added to food by processors and distributors will tend to rise while that added by producers remains stationary. If, therefore, agriculture is to maintain or increase its share of the final value of the food marketed, it must be prepared not only to become more involved in and undertake more of the marketing functions, but also to perform these functions at least as efficiently as those who are currently responsible for them.

There is a constant desire on the part of agriculture to undertake vertical integration towards the market so as to be able to improve farmers *selling power in respect of their products* in the face of greater concentration of power in the market for their products. The Food Manufacturers Federation leaves no doubt that this concentration has occurred in the food industry. According to the notes it publishes, much of the production is concentrated in complex company groups (often with international connections). 'The large-scale unit, in fact, is far more characteristic of the British Food Industry than it is of any other country outside the USA. In virtually every sector of the Food Industry, a handful or so of firms can be regarded as "dominant" in national distribution.' This claim is supported by evidence from the Census of Distribution which shows that between 1966 and 1971 the number of retail food establishments fell by 11·5 per cent and the market share of the multiples rose from 34·5 per cent to nearly 40 per cent. According to the *Annual Review of Agriculture 1974*, the total value of UK farm production in 1972–3 was divided among the different sectors as follows: farm crops 18 per cent, horticulture 12 per cent, livestock 41 per cent, milk and milk products 22 per cent, other livestock products 7 per cent. Agricultural cooperatives have at present almost no part in the disposal of milk and milk products (except in Northern Ireland) and hops, which are under the control of statutory marketing boards, or of sugar beet, where the sole buyer is the British Sugar Corporation; these commodities account for nearly a quarter of the value of fotal UK farm production. For the rest, the opportunities for cooperative enterprise differ markedly according to the commodity concerned, as will appear in a later chapter.

The financial factors affecting agricultural cooperatives are internal to themselves rather than external and will consequently be more appropriately discussed in a later chapter. However, there are several points which may be mentioned here. One of these is the *system of land tenure.* In 1972 the percentage of the agricultural acreage rented by farmers was 46 per cent in England and Wales, 43 per cent in Scotland and negligible in Northern Ireland. The percentage of rented land has been falling steadily for half a century or more; in England and Wales in 1913 the proportion was 90 per cent. In the past the predominance of rented land was a contributory factor in determining in what way the credit institutions for agriculture and agricultural cooperation would develop. Today there is, in general, no important difference so far as agricultural cooperative trading is concerned between owner occupiers and tenant farmers, but there is rather less scope for cooperation among the latter for purposes of production.

Another factor, which stands out much more prominently when one views the whole scene of European agricultural cooperation and does not consider only developments which have taken place in the United Kingdom, is the absence of any specialist *credit institutions*, either created by the State, or by the agricultural industry, or by the two in combination, for the financing of agriculture and the development of agricultural cooperation. The growth of such credit organizations, first at local level and later regionally and centrally, has kept pace in continental Europe with the growth of the local and, later, the federated cooperatives; the results of this alliance have been first a thorough understanding by the financial institutions concerned of the special needs of agriculture and agricultural cooperation, secondly a willingness on their part to give more extensive and extended credit than straightforward commercial banking considerations might have held to be justified, and thirdly an invaluable channelling of funds for investment from the non-agricultural into the agricultural sector. The very success of the banks, which has brought them an increasing number of clients who are no longer content to lend but who also wish to borrow, has steadily whittled away the former special privileges of the agricultural borrowers, so that there is now, other conditions being equal, less benefit to be had from dealing with their own banks than there used to be, though even today the knowledge of their support makes it possible for the continental cooperative entrepreneur to 'think big' in terms of future development in a way that his British counterpart would hardly dare to do.

Last of the external factors to be mentioned will be the overall *influence of government on agriculture*, which may cause it to give more attention to agricultural cooperation at one period of history than at another. Here it is reasonable to start with the assumption that a government, as do the majority of farmers, sees cooperation not as an end in itself but as one of many methods by which a chosen aim is to be accomplished. During the thirties, cooperation was considered as a method of preventing farmers undercutting one another to market their products, but rejected in favour of statutory marketing boards; the intention then was, as is shown by the wording of the Marketing Acts and the powers given to the Boards, that voluntary cooperation should be developed alongside the statutory Boards, but these turned out in practice to be irreconcilable objectives. The 1947 Agriculture Act gave producers a new type of protection in the form of a guaranteed price made up by deficiency payments for most of the major commodities (wheat, barley, potatoes, pigmeat, fat cattle, milk, etc.), this being the price which in the opinion of the government was sufficient to provide the producer with a reasonable return. These arrangements did not deprive the producer of all incentive to obtain for himself a better price than the guaranteed one, but they may well have been a factor in dulling his determination to organize for himself a cooperative marketing system under which the return from the market could be better safeguarded. Moreover, so long as there appeared to be a possibility of relying upon statutory Boards, involving no personal effort, expense, or commitment, there were few farmers who were prepared to take the trouble and experience the inevitable disappointments of organizing voluntary cooperatives to take charge of their produce.

However, an important change came about in the later sixties, with the re-emergence of international surpluses of food, the growing doubts whether statutory boards alone could meet the requirements of modern marketing, the concern felt by the industry about vertical integration and the prospect, at first obscure but later becoming more certain, of joining the Common Market. The price of entry to this was acceptance of the Common Agricultural Policy. This too provided producers with a basic guarantee, through a system of intervention prices and support buying, but one calculated to be less effective than the system they had hitherto enjoyed. British farmers were warned therefore, in the government White Paper setting out the terms of admission to the EEC, that as deficiency payments were phased out they would increasingly get their returns from the market. All of these factors tended towards a greater emphasis on cooperation, particularly in the marketing field, as a method of overcoming the problems of organization and of carrying out the revised objectives of the agricultural industry.

Increasingly, as the influence of the Common Agricultural Policy has made itself felt, it becomes necessary to examine on a wider front what is the *EEC policy towards agricultural cooperation.* At a comparatively early stage in its history (30.6.1960), in a document concerning the working out and putting into effect of the common policy, the role of agricultural cooperatives was defined by the Commission as follows: 'An effective cooperative organization can contribute in a wide sense to the improvement of the conditions of production and income, of work and of life in agriculture. By mutual cooperation in the field of supplies, processing and sales, as well as in the service sector, it is possible to improve the competitive capacity of agriculture, especially of family farming units. Cooperative development and the intervention by the right kind of cooperatives can favourably influence the vertical integration of agriculture.' However, there have been few practical steps taken to secure such development, due in part no doubt to the extreme difficulties experienced by the Commission during its first fifteen years in obtaining agreement to any general policy for the reform of agricultural structures, and its consequent desire to avoid any issues which appeared dangerously contentious. In so far as any consistent policy can be detected behind the legislation which the Commission has succeeded so far in carrying into operation, it seems to be this. The vital considerations to be kept in view are Article 39 of the Treaty of Rome, with its statement of the multiple objectives of the common agricultural policy, and the series of Articles in the same Treaty dealing with the rules of competition. In conformity with these, the Commission has found it possible to recommend a series of measures having the broad objective of grouping farmers in organizations which require their members to conform to common disciplines, and of enabling these organizations to form associations with similar conditions. The Commission's further intention is that common measures of such a kind, applying to all member States, would make it impossible for any one of them to give direct aid to its own groupings of farmers which puts them in a more favourable position than producers in another member State. The point of interest to emerge from this analysis is that EEC policy is concerned with groups, not with cooperative groups, but if it turns out that such groups are to consist solely of producers, as the Commission appears to intend, the likelihood seems to be that, in practice, most of the groups so formed will be of a cooperative character.

NOTE: There exists a considerable literature on agricultural cooperation in the European Communities (and in Western Europe generally). Useful texts, in English, are the following: On the general position in the 'Six', *Agricultural Cooperation in the EEC* (No. 21 in the Agricultural Series), EEC, Brussels 1967; on cooperative constitutions, *A Comparative Study of the Law of Agricultural Cooperatives in Europe,* Plunkett Foundation OP No. 28, 1963; on organization and trade, *The Agricultural Cooperation in the European Economic Community,* COGECA, Brussels 1974.

2. Historical summary

In contrast with other countries, where the pioneers of cooperation were agricultural producers, who developed types of it adapted to their own special needs, the first agricultural cooperative in the United Kingdom (the Agricultural and Horticultural Association, founded in 1867) arrived on the scene long after cooperatives formed by consumers, and made use of a form of cooperative legislation already well established. The Act in question was the Industrial and Provident Societies Act, a title that has mystified many generations of cooperators, and was explained by a former Registrar as follows: it was intended, he wrote, 'to intimate the design that societies should be industrial as making their profits by the mutual personal exertions of their members and provident as distributing their profits by way of provision for the future'. As a description of cooperative aims and methods this statement is less than comprehensive, and a better one will have to be attempted in the next chapter.

The most important feature of the I & P Act, for the first generation of societies and their successors, was that these societies were given incorporation with limited liability; even today, if a cooperative wished to be incorporated with unlimited liability it would have to register itself under the Companies Act. The amount of money which a member might invest in a cooperative society, and which represented his entire obligation towards it, was also limited. To understand the importance of the principle of limited liability one must remember that, in the middle of the nineteenth century when it was introduced, there was still a popular prejudice against limitation of liability where private companies were concerned, and very many disastrous failures had resulted from this. The danger was avoided by the cooperatives, which flourished under the dispensation given to them. On the other hand, the statutory limitation of members' liability towards their cooperatives has had the consequence that members, not being personally implicated in the success or failure of the trading body set up by them, have tended to regard it much as they would any other business. The problem of maintaining the personal involvement of cooperative members in their own enterprise has been a problem for cooperatives everywhere, but nowhere more than in the United Kingdom, by reason of its special legislative traditions.

FORMATION OF THE CENTRAL ORGANIZATIONS

The early history of agricultural cooperatives can be found in various publications mentioned at the end of this chapter, the main source being the records of the various central organizations. The first of these organizations to be set up in the United Kingdom, as constituted at that time, was the Irish Agricultural Organisation Society, in 1894 (from which the Ulster Agricultural Organisation Society subsequently emerged as a separate organization in 1922). The formation of the IAOS led to the establishment of an Agricultural Organisation Society, for England and Wales, in 1900 (from which the Welsh Agricultural Organisation Society separated itself, also in 1922), and to the establishment of a Scottish Agricultural Organisation Society in 1905. These organizations owed a great deal to the imagination and enterprise of Sir Horace Plunkett, an Irish landowner who in 1919 created the trust instituting the educational and research foundation which now bears his name. The first cooperatives were concerned principally with the supply of requisites, except in Ireland, including Northern Ireland, where even at the outset the dairy and agricultural cooperatives were more important. In 1909 the Cooperative Wholesale Society, a federal formed by the consumer cooperatives in 1867, was invited by the agricultural cooperatives to establish a department for the marketing of agricultural produce. The Society did not do this, but did set up an agricultural department for the wholesale supply of feedingstuffs, fertilizers and other farm requirements. Most if not all of the agricultural requisite societies in England and Wales became affiliated to and traded with the CWS (those north of the border had a somewhat less happy relationship with the Scottish CWS, which was prepared to do business outside the cooperatives) and there was also a fair amount of trade between producer marketing societies and individual consumer societies in agricultural and horticultural produce.

While the agricultural cooperatives traded on satisfactory terms with the CWS, they were few in number, compared with the consumer societies, and so could have little say in its policies. It seems likely that their objection to this inferiority was one of the principal reasons that induced the AOS leaders in 1912 to establish a Farmers Central Trading Board, which in 1918 was converted into an Agricultural Wholesale Society. This federal organization – the first to be formed by agricultural cooperatives in the UK – started life under every conceivable disadvantage. It had a well-established rival, already enjoying the support of the members it wished to attract, it was inadequately financed and incompetently managed, and it began trading at the peak of market prices which, after the end of the First World War, fell rapidly and catastrophically. The final result was that in 1924 the AWS went into liquidation and had to be wound up. The effect of this disaster on the subsequent history of agricultural cooperation in the United Kingdom (even though Scotland and Northern Ireland were only indirectly affected) can hardly be exaggerated. In the first place, many agricultural cooperatives were brought to the brink of ruin, and were saved only by the generous policy of the CWS, their principal creditor. Secondly, the principle of federal trading activity, or cooperation between cooperatives, became completely discredited, and remained so for as long as memories survived of the earlier failure, that is to say until more than a quarter of a century later. Thirdly, the Agricultural Organisation Society, which had backed the AWS, had been so much involved in its failure that it too was forced to close its doors very soon afterwards. This meant that, for the next twenty years or more, the agricultural cooperatives in England were left without any leader or spokesman for their special interests, although the National Farmers Union, by setting up a

Cooperation Committee, did something to fill the gap created by the loss of the AOS. It was particularly unfortunate that this period, when the cooperatives were without a voice of their own, should also have been one in which there were being formulated new policies for agricultural marketing, out of which emerged the Agricultural Marketing Acts of 1931 and 1933 and the establishment of statutory marketing boards. At that time it would indeed have been absurd to think of cooperatives as an alternative to marketing boards, but they should not have been allowed to disappear, as did almost all the cooperative dairies except, as already mentioned, the creameries of Northern Ireland and a very few others in the remoter parts of Wales and England.

EVENTS FOLLOWING WORLD WAR II

In England a central cooperative body was eventually reestablished, not without some protest from the NFU, in 1946, under the title of the Agricultural Cooperative Association. In the meanwhile the other national bodies had had great difficulty in maintaining their existence, and would not have been able to do so without outside help. This was most generous in the case of Scotland, where the SAOS was given special responsibilities towards the crofting population and derived much of its income from sources other than cooperative subscriptions. A rather similar situation existed in Northern Ireland, where the UAOS for most of its career has had to depend on financial help from the government. In Wales, where the subscriptions of the cooperative societies were altogether inadequate and no assistance was, at first, to be had through official channels, the situation was saved by A. W. Ashby, at that time adviser to the new Department of Agricultural Economics at Aberystwyth, through a 'Joint Scheme' initiated in 1926, under which Ashby's department made itself responsible for most of the work which the affiliated societies expected of the WAOS, in return for an annual payment which represented the salary of one assistant on the department's staff. In 1949 the ACA, SAOS, UAOS and WAOS established, together with the Irish Agricultural Organisation Society, Agricultural Cooperative Managers Association and Plunkett Foundation, a Federation of Agricultural Cooperatives which, though it did not have any executive role, for many years performed a valuable service in bringing the various central organizations into a consultative relationship, which helped to pave the way for the developments that were to follow.

The year of the Federation's formation, in the aftermath of World War II, but while a great number of the controls on normal commercial competition still operated, may be a good one in which to take a first view of the general shape and size of the British agricultural cooperatives, which at that date numbered 402 societies with an annual turnover of £48·6 million, 52 per cent of which was accounted for by requisites and services, and 48 per cent by marketing, the latter consisting mainly of grain, eggs and horticultural products, with some livestock, meat and wool. Following the relaxation of wartime controls the cooperatives began once more to enlarge their membership, which indeed continued to grow until the early sixties, when it reached a peak and began to fall again. (The number of agricultural holdings had been declining since the early fifties.) Membership carried no obligations beyond the subscription of a modest number of shares and while a loyal few put the bulk of their business through their cooperative, the majority of farmer members used it only occasionally or as a trader of last resort. This was the beginning of the era of the take-over or, in cooperative parlance, 'transfer of engagements' from the weaker societies to the stronger ones. A precisely similar trend was taking place in the private sector, but was taken there to further extremes, with the result that by the end of the sixties the farmer cooperatives, which had earlier been numbered among the larger organizations in the agricultural merchanting sector, found themselves outclassed in size by much bigger amalgamations of private concerns. It was at this stage that many farmers, becoming concerned about the rapid decline in the number of trading sources and outlets remaining open to them, began to urge that a new look should be taken at cooperation, its problems and opportunities.

The national leadership of agricultural cooperation was less effective than it might have been at this time, because it was so often in dispute. The troubles of the central organization for England – always the most important of the four – were by no means over when it re-established itself in 1946, since within a few years it was confronted by a rival NFU backed central organization. This organization amalgamated in 1956 with ACA, which then changed its name to ACCA (Agricultural Central Cooperative Association). Ten years later, in 1966, ACCA changed back into ACA again, on the withdrawal of the NFU Development Company as a shareholder. Finally, in 1972, ACA was converted into ACMS (Agricultural Cooperation and Marketing Services Limited) in which the NFU once more has an interest. Behind these political manoeuvrings there lay genuine ideological differences, though with the passage of time they have lost their former intensity. They revolved round a central, much debated question, which had to be settled before the future course of agricultural cooperative development could be determined. (An attempt in 1961 to settle it by forming a common federal – Agricultural Central Trading – proved quite abortive.) The question being asked was whether cooperatives should continue to provide what was basically the same service to all their members, or should discriminate in favour of members who, by grouping together, calculated that they were able to save overhead costs, from which saving to the society they as a group ought to be able to benefit. In practice the extent of the saving was often overestimated by the group, and the cooperatives were reluctant to make concessions which might have repercussions on their other members, with the result that the groups often ended by setting themselves up independently, and regarding themselves as a distinct kind of cooperative. Many of them became incorporated under the Companies Act, sometimes because they had fewer than the seven members required under the Industrial and Provident Societies Act, but more often to emphasize that they considered themselves a different sort of organization.

POST-WAR GOVERNMENT POLICIES

To pick up the sequence of events in this history it is necessary to go back to 1954, the year in which meat was de-rationed. Thereafter it began to appear, increasingly during the later nineteen fifties, that the pre-war pattern of surpluses could again be expected and that the emphasis in agricultural policy, which had so long centred exclusively on production, would have to be broadened again to cover marketing. In 1956 the British Egg Marketing Board was established, but this was the last statutory marketing body to be set up, and the policy-makers began looking for alternatives, in particular at the possibility of encouraging voluntary cooperation. The first tentative steps in this direction were taken in the Horticulture Act 1960, one short clause in which dealt with grants for promoting cooperative horticultural marketing. A more fundamental measure, which has been called the charter of agricultural cooperation, though at the time few recognized it as such, was the Agricultural and Forestry Associations Act 1962, which removed some of the more particularly vulnerable agreements entered into by agricultural, horticultural and forestry cooperatives from the scope of the Restrictive Trade Practices Act 1956. In order to provide these cooperatives with this exemption, it became necessary for the 1962 Act to define what constituted an agricultural (or forestry) association; the three criteria adopted were, first, that the association could be registered under either the Industrial and Provident Act or under the Companies Act but in the latter case must be substantially cooperative in character, second, that at least ninety per cent of of the voting power had to be attached to shares held by persons occupying land for agricultural (or forestry) purposes, and third, that the only or principal business must, where agricultural operations were concerned, be the marketing of produce or the supply of goods required for production purposes. Although this definition was invented to meet a particular situation, namely the problem being dealt with under the Act, it tended to be followed in other circumstances where agricultural cooperatives were concerned. Two years later, a fourth criterion appeared in the Agriculture and Horticulture Act 1964, which extended the range of grants given under the 1960 Act but laid down that aid would not be given to cooperatives unless the appropriate Minister could be satisfied that the constitution contained a provision to ensure that the members showed a degree – not yet defined – of loyalty towards it. This provision marked a decisive point of separation between the old philosopy of cooperation, in which the cooperative was regarded as having no 'rights' of its own, to the new one, in which it was appreciated that the members had to give 'rights' to the cooperative, surrendering some of their own, in order to ensure that it would be able to carry out the tasks assigned to it.

The next date in this historical review, 1967, has been taken for two reasons. In the first place it marked the centenary of the establishment of the first agricultural cooperative in 1867. By 1967 (more precisely, in the year ended 31.3.1968) there were 540 of these cooperatives – not counting the 600 or more small pest-clearance societies which, though they performed a valuable task, were statistically insignificant. Their trade amounted to £296 million, of which some 48 per cent was attributable to marketing and the balance to supply and servicing activities. More than two-thirds of this turnover originated in England, after which came Scotland with approximately a fifth, next Wales, and lastly Northern Ireland. Members' shareholding stood at £21·5 million, and reserves and undistributed surplus at £11·1 million, making a total owned capital of £32·6 million. What these global figures do not reveal is the extent to which the cooperative sector was dominated by the larger societies, but as the same point will be discussed when more recent figures are examined, it need not be entered into here. The cooperatives, like agricultural businesses generally, were going through a difficult period; their average net surplus in 1967 was 1·8 per cent, out of which they had to pay a relatively high interest on shares, appropriate a sufficient amount to reserves to take care of inflation, and find enough by way of bonus to keep their trading members content. Few were able to accomplish all these objectives, and all were deeply concerned about their future.

PRELUDE TO THE 1967 ACT

The year 1967 was also significant in agricultural cooperative history for a second reason, that it saw the passing of an Agriculture Act, Part IV of which was solely concerned with agricultural cooperative development. Before going into this, it is worth mentioning that the three Acts previously noted, those of 1960, 1962 and 1964, had all been passed by a Conservative administration (the 1962 Act being technically a Private Member's measure, but one which the government of the day had fully supported). In 1965 a Labour government came into power. It too pursued the aim of encouraging cooperation but chose a different method of doing so. For the first time specific objectives were declared, not only of where and why the various aids would be given, but also of how the measures proposed for cooperation would be related to other measures dealing with other aspects of agricultural structure. The total package, in the form of a government White Paper with the title *The Development of Agriculture* (Cmnd 2738), appeared in 1965 and contained four parts, headed, respectively, 'Farm Structure', 'Farm Improvement Scheme', 'Hills and Uplands' and 'Cooperation'. The first and fourth of these headings were clearly linked in the introduction to the White Paper which envisaged various ways of helping, in particular, the 100 000 or more small farmers with businesses of less than 600 standard man days (the upper limit of the revised Small Farmers Scheme) who depended upon farming for all or most of their living. The courses of action suggested for them were to enlarge their farms where they could get more land, to cooperate with others to have some of the benefit of farming or marketing on a bigger scale, or to retire from farming where they wanted to give up an unrewarding struggle. In passing, it may be noted how nearly this concept was related to the Mansholt proposals in the EEC plan *Agriculture 1980* which was to appear a year or so later, and how closely in principle the idea of a 'middle band' of farmers to be given special assistance resembles that later embodied in Directive EEC/72/159, applying to all EEC member States a common agricultural policy in respect of aids for agricultural production.

Returning to the White Paper, the section dealing with cooperation reiterated the special need for cooperation among small farmers, the point being made no less than three times in the opening paragraph of this section. There is thus no doubt about its principal intention, which was not a new one. In other respects, however, the White Paper broke fresh ground. In the first place it strengthened the criterion first enunciated in the 1964 Act; a cooperative must not only possess, but must also make use of, powers to ensure that members fulfilled certain minimum obligations and were loyal to it. Secondly, it set out to give substantial encouragement to cooperation in production, as well as in marketing, this form of cooperation being seen mainly as an alternative to amalgamation, though sometimes also as leading to it. Thirdly, it recognized the importance of preliminary studies and surveys, to ensure so far as possible that farmers' and government money was wisely spent; such studies were to be given generous encouragement. Fourthly, it proposed an additional function of discovering and popularizing new forms of cooperative activity among farmers. Fifthly, it envisaged the setting up of a new agency, to be called the Central Council for Agricultural and Horticultural Co-operation, which would be given responsibility for carrying out the government's plans and allowed discretion in doing so; the Council was to be guided on the one hand by the degree of commercial and technical merit in the proposals put to it, and on the other by the status and resources of the recipients. We return to the subject of the Central Council in Chapter 10. Here it need only be added that, because of the intervention of a general election, the legislation needed to carry out the government's intentions was not completed in the year following the publication of the White Paper, but came into force in the Agriculture Act 1967.

The Development of Agriculture was thoroughly innovatory and on the whole well inspired. It rested, however, on certain assumptions, one of which in particular needs to be further examined. The authors of the White Paper, like many before them, assumed that there was, or if not that there ought to be, some connection between cooperation and the smaller farmers. This is a subject about which too little is known, but it must be mentioned that in a study carried out by the A. C. Nielsen Company Ltd in 1963 the positive correlation between membership of a cooperative organization and size of holding rises with the size of holding, and is therefore lowest where small farmers are concerned. Too much weight should not be given to this finding, though it is borne out by experience that the persons most interested in cooperation tend to be better educated, more progressive and, on the whole, larger farmers. A cooperative formed to service small farmers alone would be a doubtful economic proposition, whereas a cooperative with a mixed membership of small and large farmers can be economically viable and give to all its members a service corresponding to their needs.

The question of the relevance of government aid to agricultural cooperatives, the various forms it may take and the results it is able to achieve can be deferred until later. Here it is only necessary to make the general statement, in justification of the amount of space which has been devoted to government policy in a chapter dealing with the historical background of agricultural cooperation, that the general attitude of government in relation to agricultural cooperation is important in itself, and at least as important as the aid which it actually makes available. In support of this claim one need only point to the fact that, in a country so dependent on agricultural imports as the United Kingdom has always been, government policies are necessarily of fundamental concern to the farmer, to whom they can make the difference between prosperity and penury. Seen in historical perspective, therefore, the interest which has begun to be taken in agricultural cooperation since the last war, first by one political party, then by the other, and which now seems to represent common ground between all parties, is a historical fact of the first importance, which certainly does not explain the agricultural cooperative revival which took place during the post World War II period, but helps to account for its continuing momentum.

POLICY AND PROGRESS IN THE SEVENTIES

The next stage in the development of government policy towards agricultural cooperation may be said to have begun, innocently enough, with the appointment by agricultural Ministers in 1971 of a committee to assemble and consider the information about existing arrangements for the production of crops, etc., under commercial contract, and to advise on the extent and type of likely future developments in this field. When they introduced their report in the following year, the Committee acknowledged that, encouraged by the *Annual Review* White Paper of March 1972 (Cmnd 4928), they had made a wider ranging study of agricultural marketing than their terms of reference might have allowed. This 'Barker report' (*Contract Farming* – Cmnd 5099) had great significance for agricultural cooperation not only for what it said, which was a good deal, but even more because the body saying it was an independent committee, whose impartial verdict on the role that agricultural cooperation would be called on to play in the future development of an advanced and sophisticated agricultural system obviously carried more weight than any similar claim which the cooperatives might think of making for themselves. It is necessary to explain the Committee's allusion to the *Annual Review* White Paper. In this White Paper the government, recognizing the particular importance that efficient marketing would have after accession to the EEC, had invited the Farmers' Unions, the Central Council and other appropriate organizations to consider how voluntary cooperation for marketing purposes could be further encouraged and how producer marketing organizations could be best developed. The government's tentative conclusions from the various submissions which resulted from this invitation, including the report from the Contract Farming Committee, were embodied in a Green Paper entitled *Agricultural and Horticultural Marketing* published in June 1972 (Cmnd 5121). This was a consultative document which, however, accepted the Committee's judgement of contract farming, and that producer groups and second tier groups needed developing. The Green Paper again invited views, and proposed a further round of consultations with the principal trade associations and bodies concerned. The chief of these bodies was a consortium of political and commerical organizations known as the British Agricultural Marketing Development Organisation (BAMDO) – consisting of the three Farmers' Unions, the five UK Milk Marketing Boards, four other Marketing Boards, the four Cooperative central organizations, and the British Farm Produce Council – which was formed in the first instance in order to enable its various constituent producer-controlled bodies to coordinate their views, and to act as a joint negotiator on their behalf. Thus by July 1973 the position had been reached where there was, on the one hand, a government-sponsored body, the Central Council, with responsibilities of a public character, and, on the other, a broadly based body of a private character, namely BAMDO. The relationship subsequently worked out between these two bodies and the details of the measures adopted by the government for the further encouragement of agricultural cooperation, based on this new structural framework, can be more conveniently dealt with later, in Chapter 10.

In Chapters 7–9 an examination will be made, sector by sector, of the agricultural cooperatives in whose interests all this political activity has taken place. It will, however, be convenient to close the present chapter, concerning the historical past, by presenting an up-to-date view of where these cooperatives now stand, taking for this purpose the statistics for the year 1973, the latest that are available. The information may be more easily digestible if it is confined to main facts presented in a tabular form, in which a division is made at some points between the separately collected figures of cooperative societies registered under the Industrial and Provident Societies Act and of cooperative-type companies registered under the Companies Act, although both are concerned with the same kind of cooperative activity.

		SOCIETIES		COMPANIES
(a) Total turnover		£557 million		£139 million
(b) Proportion of (a)	in England	66%		99%
	in Wales	6%		
	in Scotland	22%		1% (Wales, Scotland, N Ireland)
	in N Ireland	6%		
(c) Proportion of (a)	in supplies	70%		17%
	in marketing	30%		55%
	in services	–		28%
(d) Estimated % share in	feedstuffs supplied		15%	
	fertilisers supplied		18%	
	cereals* sold		12%	
	potatoes sold		9%	
	fatstock (1971) sold		8%	
	eggs sold		17%	
	top fruit sold		20%	
	open air vegetables sold		9%	
(e) Number of organizations		402		186
(f) Members ('000)		315		29
(g) Share capital		£28 million		£2 million
(h) Reserves and surplus		£22 million		£100 million†
(i) Total capital employed		£133 million		£120 million†
(j) % of (g)+(h) to (i)		38%		85%

* includes cereals bought for own manufacture.
† includes £96 million insurance cooperative reserves and surplus.

The above figures may help to convey the order of magnitude of the various problems concerning agricultural cooperation which are to be discussed in the next four chapters, and to place them in a comprehensible perspective.

NOTE: The early history of agricultural cooperation has been dealt with only sketchily in this chapter since it is already available, for England in Part I of *An Analysis of Agricultural Cooperation in England*, a report made to ACCA by Dr J. G. Knapp, 1965, for Wales in Chapter 7 *Cooperative Trading in Wales* published by the Department of Agricultural Economics, Aberystwyth, and in various chapters of *Farmers Together* edited by Elwyn Thomas of WAOS, 1972, for Scotland in a *Report of the Committee on Agricultural Cooperation in Scotland* published by HMSO in 1930 (Cmnd 3567), for Northern Ireland in *Agricultural Cooperation in Northern Ireland* by Professor Parkinson, published by HMSO in 1965 (Cmnd 484), and for the UK generally in *Agricultural Cooperation in the United Kingdom* by Margaret Digby and Sheila Gorst, 1957, to mention only a few of many sources. The various agricultural cooperative figures quoted are taken from the series *Agricultural Cooperation in the United Kingdom*, published by the Plunkett Foundation.

3. Constitution and characteristics of agricultural cooperatives

Agricultural cooperatives in the United Kingdom appear under different guises. They may be organizations incorporated under the Industrial and Provident Societies Acts 1965 to 1968 or the Companies Acts 1948 and 1967 (these Acts do not extend to Northern Ireland, which has separate but similar Acts of its own), or they may be unincorporated, in which case they fall within the scope of the Partnership Act 1890. The third of these forms is of use in a rather limited range of circumstances and will warrant only a brief description.

AGRICULTURAL COOPERATIVE PARTNERSHIPS

There are three considerations limiting the usefulness of cooperative partnerships. The first is, in the words of the Companies Act 1948, that 'no company, association, or partnership consisting of more than twenty persons shall be formed for the purpose of carrying on any business . . . that has for its object the acquisition of gain by the company, association, or partnership, or by the individual members thereof', unless it is registered as a company (or a cooperative). The second is that each and every member of a partnership is normally liable for all the debts of the firm, whereas the liability of a member of a registered cooperative or company is limited to the amount of his shareholding. The third is that a partnership (in all parts of the United Kingdom except

Scotland) has no legal personality distinct from that of the partners of which it is composed. These three considerations, particularly the second one, make a partnership an unsuitable body to undertake any trading functions involving risks which are outside the members' control – though it may be noted, in passing, that unlimited liability was quite normally accepted by the farmer members of continental cooperatives, at an earlier stage of their development. On the other hand, a partnership is an entirely appropriate form of organization for a small group of farmers who wish to cooperate for some purpose which is primarily concerned with production, and thus under their direct supervision. The responsibilities and rights of the partners will usually be defined by a partnership agreement. The question whether the terms of the agreement are such as to make the partnership a cooperative one has become less important than it used to be, now that the grants for the modernization of farms which are available under EEC legislation (Directive 72/159) can be made to groups of farmers generally, whether or not they are organized on a cooperative basis.

AGRICULTURAL COOPERATIVE SOCIETIES

These are agricultural cooperatives which are incorporated and registered under the Industrial and Provident Societies Act 1965. This Act is a general one for cooperative societies of all types, consumer cooperatives and productive (i.e. manufacturing) cooperatives as well as agricultural cooperatives, and it recognizes no difference between them, except in one section of minor importance. Consequently, in order to find a precise distinction between an agricultural cooperative and one of another type, it is necessary to look elsewhere than in this Act. This question will be considered presently.

Far more important, because it is a source of continual confusion and misunderstanding, is the difference between a cooperative and an ordinary company. Superficially, there is no difference. Cooperatives and companies are both incorporated organizations, with limited liability. Both have shareholders, who in a cooperative are more commonly referred to as members. Both aim to achieve a profit, in a cooperative more accurately described as a surplus, on their trading operations. Nevertheless, cooperatives and companies are concerned with fundamentally different objectives. Unless this difference is properly understood cooperatives will be set tasks which it is not appropriate that they should undertake, or will be judged by criteria which are not properly applicable to their circumstances.

The standards to which cooperatives are expected to conform are often referred to as cooperative principles, inaccurately, and perhaps unfortunately, seeing that the ordinary person tends to be more concerned with the practice of what he does than with the principle of doing it. As a result of the Prevention of Fraud (Investments) Act 1939, which laid down that in future a society could only be registered as such if it was a bona fide cooperative society, it became necessary for the government to set out with some precision what were the objects such a society must have, and how it would be expected to conduct itself, in order that recognition might be given to it. The criteria established are no less valid now than formerly, as will be apparent from the following quotations:

> an investment Society . . . is expressly excluded, i.e. a Society which is carried on with the object of making profits mainly for the payment of interest on money invested with or through the Society.

It was evidently thought useful to state what a socitey is *not*, before stating what it is. The language used here has been carefully chosen to make plain the contrast with investment in an ordinary company, the object of which is, quite simply, to obtain in the short or long term the best possible return on the money invested; indeed it can be fairly said that the shareholders in an ordinary company have in general no other interest than this.

The converse is found in the statement of what *is* the purpose of a cooperative society:

> The Society must so conduct its business as to show that its main purpose is the mutual benefit of its members, and that the benefit enjoyed by a member depends upon the use which he makes of the facilities provided by the Society and not upon the amount of money which he invests in the Society.

Two points are being made here. The first part states that the main purpose of the Society is the mutual benefit of its members, that is to say if they are consumers by furnishing them with supplies of good quality at lower cost, if they are producers by providing them with secure outlets where they can earn a premium, but in either case mutually, since the benefits which can only be obtained by collective action have also to be equitably divided. The second part takes up the point made by the first, and elaborates it in a phrase which is perhaps more frequently quoted than any other to illustrate the essential difference between a cooperative's and an ordinary company's method of operation; it says that the benefit to be enjoyed by a member depends on his use of the Society's facilities. To make this second point absolutely clear, the paragraph concludes by taking the example of a society such as an agricultural cooperative society, where

> although the member may be required to take up shares in proportion to the amount of his land or stock, etc., the Society nevertheless exists primarily to provide benefits for the member in proportion to the use which he makes of the marketing or other facilities furnished by the Society.

It is to be noted in particular that the statement refers to benefits, not to profits. The highest benefit that a society could obtain for its members might very well be to give them trading terms close to the actual cost of the operation. Indeed, a number of cooperatives have set out to do exactly this, though most of them find it not only safer but also advantageous from other points of view to realize a surplus on their year's trading which can be retained in the business rather than divided among the members, if this course seems advisable. However to aim at a surplus for reasons of commercial prudence is quite different from taking profit to be a main objective.

Finally, in order to ensure that the primary benefits flow to the member in proportion to his business activity, rather than as a consequence of his investment, the memorandum of guidance issued by the Registrar of Friendly Societies (from which this and the previous quotations are taken) contains the following direction:

> The return on share and other capital must not exceed a moderate rate which may vary according to circumstances but should be approximate to the minimum necessary to obtain such capital as is required to carry out the primary objects of the Society.

As to what constitutes 'a moderate rate' there have been differences of opinion and practice in recent years, under the influence of growing inflation, but the guiding principle has remained unaffected. In conjunction with other provisions relating to distribution of benefits, this rule of limited interest on capital illustrates the profound difference between a cooperative and a non-cooperative type of organization.

AGRICULTURAL COOPERATIVE COMPANIES

Companies may, however, also be constituted so as to provide for the principal benefit from their operations to be passed to members in proportion to the use made by them of the company's services, and for the return on the shares to be limited to a rate which is sufficient to obtain the necessary capital. During the past decade many companies have been established by agricultural or horticultural producers which fully satisfy this main cooperative criterion, and other subsidiary criteria to be mentioned later in this chapter. As will be seen, there are some special circumstances which favour the formation of a cooperative society under the Industrial and Provident Societies Act, and others which favour the formation of a cooperative-type company under the Companies Act, though it must be added that it is just as often the personal preference (or prejudice) of the cooperators (or their advisers) which determines the choice of constitution. The growth in the number of agricultural cooperatives registered under the Companies Act has led to some administrative difficulties, in two directions. First, the Registrar of Companies is not the least concerned whether a company in his registry is cooperatively constituted, or conducts itself according to cooperative principles. His responsibilities cannot be compared with those of the Registrar of Friendly Societies (or Registrar of Companies and Friendly Societies in Northern Ireland) whose official duty it is to ensure that the rules of any cooperative association submitted to him for registration are of an appropriate character, that any subsequent changes of rules do not destroy this character, and that the association's conduct is in accordance with its rules; this last fact being ascertained from the annual return which, under the law, each cooperative society is required to make to him. In consequence, there is no automatic mechanism for determining which companies are cooperative, or whether a company claiming to be a cooperative is so in reality. Secondly, agricultural cooperatives have had many occasions in recent years to press for amendments to the law which would take account of their particular situation. On some of these occasions they have been successful, but they would certainly have been able to obtain the results with less effort, and might have achieved better results, if their demands had not involved two differently constituted, though in most other respects entirely similar, types of organization. The reader will notice, in this and the next chapter. the extent to which the description of the contemporary agricultural cooperative system has been complicated by the need to take both societies and companies into account.

COOPERATIVE CHARACTERISTICS

Throughout their history cooperatives have, understandably, devoted much time and thought to the aims which motivate cooperative activity, and to considering the characteristics which should distinguish a cooperative from other types of enterprise. In 1966 a Commission on Cooperative Principles reported to the International Cooperative Alliance, dealing at length with the questions of Disposal of Surplus and its corollary, Interest on Capital, which have already been discussed. Other cooperative principles mentioned in the ICA report, which are reflected to some degree in United Kingdom practice, were the following:

(i) Membership of a cooperative, in the ICA recommendation, 'should be voluntary and available without restriction or any social, political, racial or religious discrimination, to all persons who can make use of its services and are willing to accept the responsibilities of membership'. Obviously, the voluntary aspect of membership is a fundamental characteristic of cooperation, and is what principally distinguishes an agricultural cooperative from a marketing board. The principle of general availability of membership, or 'open door' policy as it is sometimes called, is more contentious. It is, to an extent, accepted as a principle applying to United Kingdom cooperatives; the government memorandum previously quoted having the following to say on the subject

> There must be no artificial restriction of membership with the object of increasing the value of proprietary rights or interests. On the other hand there may be reasons for restricting membership which would not offend the cooperative principle, e.g. . . . a Society may confine its activities to a particular class of persons or to a particular area. By contrast, if the membership were limited in order to give the maximum benefit to a restricted number of persons the Society might not be regarded as truly cooperative.

Of this one can only say that it is usually easier for the cooperative to find reasons for restricting membership (second sentence) than for any outsider to be able to prove that membership has been restricted for the wrong reasons (first and third sentences). Such an occasion would arise, for instance, if a cooperative habitually accepted a person's trade but was unwilling to admit him as a member, even though he accepted all its conditions. In the main this principle has not given rise to much debate in the United Kingdom.

(ii) Democratic administration is another of the ICA's principles recognized by the government memorandum, which states:

> A rule providing that any persons should have more than one vote might suggest *prima facie* that the Society was not a true Cooperative Society.

In a primary cooperative it is usually considered a satisfactory application of this principle if the rules provide that no one member is able to exercise more than one-tenth of the votes in a general meeting. In a federal, consisting of two or more primary cooperatives, one of which has many more individual members than another, the arrangement may need to be more flexible, but must always have regard to the same principle, that it is the individual member whose interests have to be taken into account.

Another safeguard which is provided by the Industrial and Provident Societies Act, but for which there is no parallel in the Companies Act, is the limiting of the shareholding of an individual member (other than a member which is itself a cooperative) to a maximum amount. This maximum is currently £1000. Whatever figure is chosen is likely to be too low for some societies and too high for others; it has often been suggested that a more flexible formula should be adopted.

Finally, under the same heading may be considered the insistence of the Industrial and Provident Societies Act on a minimum of seven members, presumably based on the view that there is cooperative safety in numbers. But an agricultural cooperative, especially one concerned with production, may well find that a membership of seven would be impracticable. Such a cooperative, if it needs to become registered, must choose to operate under the Companies Act.

(iii) Other principles recognized by the ICA concerned the provision of educational facilities by cooperatives, and cooperation between cooperatives at local, national and international levels. Both of these, but particularly the latter, come more within the category of pious aspirations than within that of principles which are habitually transformed into practice.

The ICA Commission, which elaborated the principles of cooperation, and emphasized the importance of freedom to join a cooperative, also had something to say about the facility of resigning from it. The Commission recognized that in many cases the cooperative cannot allow the member to resign without his giving ample notice, and even then may have difficulty in repaying his share capital and so terminating his membership. These financial considerations will be discussed in a later chapter. But there is one further complication which must be mentioned here. This is that a cooperative registered as a company is unable to buy back its own shares, since to do so would involve a reduction of its capital without the consent of the court. A cooperative registered as a society is under no such difficulty, since its capital is freely variable. Chapter 4 mentions ways by means of which a company can overcome this difficulty, provided that it is aware of the problem, and builds appropriate safeguards into its constitution.

THE AGRICULTURAL AND FORESTRY ASSOCIATIONS ACT 1962

The discussion so far in this chapter has centred round the Industrial and Provident Societies Act and the Companies Act, neither of which makes any distinction between agricultural and non-agricultural organizations. With the developing government interest in agricultural cooperation as an instrument of public policy it clearly became necessary to have some formal understanding of what constituted an agricultural (or forestry) association and, as already mentioned in Chapter 2, this was reached by means of an Act which defined such an association in terms of its membership and its purposes, and which further ensured that these purposes would be carried out in the interests of members and not, except to a subsidiary extent, in the interests of third parties. So far as members of cooperative societies were concerned this point was already covered by the declaration that a society had to make, in its annual return to the Registrar, of trade done with non-members. The same principle was now extended to cooperative-type companies which, to qualify for exemption under the Act, must carry on their business so that 'in any period of three consecutive financial years of the association, the value of the produce bought from or marketed for persons who are not members of the association . . . shall not exceed one-third of the respective values of all the produce marketed [and] goods supplied. . . .' Here again there is an affirmation of the principle that a cooperative is in business for the purpose of providing its members with a service, not of making money out of third parties.

What made the passage of this legislation so urgent was the realization that the definition of a trade association in the Restrictive Trade Practices Act 1956, i.e. 'a body of persons . . . formed for the purpose of furthering the trade interests of its members', caught hold fairly and squarely of agricultural cooperatives. The Act stated that any agreement made by a trade association would be construed as if it had been made between all persons who were members of the association, the effect of which would have been to render cooperation among farmers practically unworkable. The Agricultural and Forestry Associations Act 1962 therefore gave agricultural cooperatives various exemptions from the Restrictive Trade Practices Act 1965 which, although subsequent experience has shown them to be insufficiently comprehensive, at least made it possible for them to continue in business.

MODERN CONCEPTS OF AGRICULTURAL COOPERATION

It is conceivable that the next major change in the legislative framework of agricultural cooperation will take place in a European context. In Chapter 12 mention will be made of the effect that the implementation of the EEC's Common Agricultural Policy has already had on the development of agricultural cooperative enterprises in the United Kingdom, and of the plans for elaborating an EEC statute for European cooperatives, a draft of which has been sent to the Commission for consideration. Here it will be appropriate to mention some of the ideas which are in circulation, both here and abroad, concerning changes that may have to be made in the law and form of organization of agricultural cooperatives if they are to continue to play a significant role in the future.

The central fact of the present situation of agricultural cooperatives is that, while their form of organization has not bascially altered during the last century, the economic environment in which it was designed to operate has changed radically in the course of the past one or two decades. The main element of change has been the emergence of large corporations, nationally or internationally based, in the agricultural supply or marketing sectors of the economy, where previously there had only been local firms, economically insignificant and usually inferior in stature to the cooperatives themselves. A second factor has been the development of agribusiness, which removes bodily out of the market a sizable proportion of the trade which agricultural cooperatives would expect to handle. (It is difficult for an agricultural cooperative to become an agribusiness itself, since by undertaking production it would enter into competition with its own members.) A third is that the function of marketing, and to hardly any lesser extent that of organizing supplies, has quite rapidly become highly capital-intensive. And, fourthly, the recent past has been a period of continuous inflation. All these factors have combined to hit agricultural cooperatives in their weakest spot, which is the ability to mobolize large amounts of capital, and to do so at short notice.

This financial problem will be the subject of the next chapter. Its legal aspect is how to obtain capital from sources outside the active membership of an agricultural cooperative without endangering farmers' control over it. Various methods of doing so have been suggested, and in some cases actually applied. In France the cooperative law has recently been changed to allow 'commanditaires', that is to say inactive members, to hold a proportion of the shares in an agricultural cooperative and to exercise a proportion of the votes. In other countries, changes in the law have not so far become necessary, mainly because of the existence of specialist banks or credit institutions which were originally farmer-owned and which still have a strong agricultural orientation. These can, for the time being, provide sufficient loans to the agricultural cooperatives in the countries concerned to enable them to make the heavy investments in the storage, transport, grading, packing, processing and refrigeration facilities, which they are being called on by their members to undertake. But the question already arises, whether these loans will not have to be replaced by a more permanent form of finance.

Whatever the ultimate form of the arrangements, they all envisage a substantial investment by persons or bodies not actively trading with the cooperative. These persons' interest must be the same as that of financiers of any other business venture, namely in a high rate of return, coupled with a capital appreciation of their investment, and the degree of control that is necessary to ensure that a proper financial policy (i.e. proper from their point of view) is carried out. Whether these aims can be accommodated in a cooperative so that a balance can be struck between them and the aims of the active users of the cooperative is a question that cannot be answered as yet with any certainty. In the United Kingdom, where there is no specialist institution for financing agricultural cooperatives, but they must borrow from commercial banks on the same basis as any other business, the question has some urgency. A partial answer to it has been found, within the existing law, permitting an agricultural cooperative society to offer preferred terms to those of its members who are no longer actively trading with it, but whose investments of capital are important, and even vital, to its financial viability. Agricultural cooperative companies, which are unable to convert ordinary shares to preferred shares, have to find a more circuitous way round this problem.

The above paragraph deals with one of the dilemmas confronting cooperatives. Another, equally painful, arises out of the growth in size of these organizations. Size is unavoidable. The commercial arguments in its favour are overwhelming. Furthermore, organizations which are unable to provide an adequate career structure for first-rate managers will fail to attract and retain them; this does not mean that they have to be mammoth corporations but equally they cannot afford to remain small. But it has to be admitted that large-sized organizations, however conscientiously they pursue the aim of good member-relations, find it difficult to be cooperative in the full sense of the word. In particular members will not readily be persuaded to provide the finance that is necessary for the services they require – probably at the cost of development on their own farms – to a remote and powerful organization, whereas they may be ready to do so where the finance is clearly intended to be spent on some cooperative facility which is visible to them and from which they will personally benefit. What seems to be needed is an arrangement whereby groups of farmers, who are prepared to contribute towards this kind of investment and to accept the disciplines which make it work, can be brought into some system of relationship with the big multi-purpose cooperatives which will, on the one hand, allow the group a considerable degree of self-government, but, on the other hand, ensure that the centralization of marketing strategy and financial policy is not undermined. Here too a partial solution can be found without going outside the existing cooperative law. The solution is to establish 'affiliated' cooperatives with limited objectives, which cannot be altered or enlarged without the consent of all the members, including the 'parent' cooperative as one of these members. In this way it is possible to achieve, within a wholly cooperative context, the same sort of relationship between a large general purpose organization and a smaller specialist one, which has been developed to such good effect by ordinary commercial and industrial companies.

To sum up, there are signs that an agricultural cooperative

philosophy, as reflected in law and forms of organization, which has been perfectly adequate for the past hundred years, when the average size of business in the agricultural sector was small, when trade was mainly local, and when prices were reasonably steady, may no longer be adequate for the future, when none of these conditions are likely to apply. In principle therefore it must be expected that both the philosophy and the law will have to change; the obvious danger is that, in the course of changing, the purposes and reasons for cooperating may undergo such transformation as to make cooperation no longer attractive to the very persons who are supposed to be its main beneficiaries. There is no need to be too pessimistic, however. While cooperatives have obvious difficulties in adjusting themselves to modern conditions, they clearly have important advantages too. The ideas which they incorporate – the emphasis placed on membership rather than shareholding, on participation rather than direction, on service rather than profit as the ultimate objective – come much nearer to the spirit of the present times than those by which many of their rivals in business are activated. It would be astonishing indeed if these advantages cannot, in one way or another, be turned to good commercial account.

NOTE: The *Report of the Working Party on Agricultural Cooperative Law*, obtainable from the Central Council for Agricultural and Horticultural Co-operation, goes in depth into many of the matters raised in this and the following chapter. Another useful reference work is the *Handbook to the Industrial and Provident Societies Act, 1965*, by W. J. Chappenden, published by the Cooperative Union Ltd, Manchester.

4. Finance and taxation

At the end of 1972 there were 9429 cooperative societies of all types registered under the Industrial and Provident Societies Act, compared with effectively 542 578 companies of all types registered under the Companies Act (excluding Northern Ireland registrations in both cases). This disparity of numbers helps to explain the general lack of understanding of how cooperatives work, and in what way they differ from ordinary companies, which exists not only among the public at large but also to a considerable extent among their professional advisers.

OWNED CAPITAL: SHARES, MEMBERS' LOANS AND RESERVES

A main cause of confusion is that the same terms are employed to describe company and cooperative concepts which are not themselves identical. Consider 'shares' for example. The shareholder in an ordinary company (i.e. not one which has been so constituted as to give it most of the characteristics of a cooperative society) is the owner of what is, first and foremost, a financial asset which – in economic theory, and to a large extent in practice – he will be prepared to exchange for an investment in any other company, if he calculates that he is receiving an insufficient return on it. Where the company concerned is a public company, he may be able to measure the value of one share against another by reference to a Stock Exchange quotation. This value can be higher or lower than the nominal value at which the share was issued, or the price at which it was acquired. Every share carries a vote, so that an investor with a large shareholding is able to exercise a greater influence over company policy than an investor with a small one. ('Money talks', as they say.) If the shareholder wishes to sell his share, he must find some purchaser or broker, other than the company itself, who is prepared to buy it. A company is not allowed to buy its own shares for the reason that, if it did so, it would be able to reduce its capital, the amount of which is stated in its Memorandum of Association, and constitutes a guarantee to the company's creditors.

In a cooperative, however, where the member expects the main benefit of membership to accrue to himself and other members in proportion to trade done with the cooperative, he must obviously accept some corresponding limitation of the rate of interest on his shareholding. Moreover, he must resign himself to the fact that, in the ordinary course of events, there will be no capital appreciation of his shares, for the reason that his only claim on the reserves of the cooperative is for a division to be made on the basis of past business done, rather than on that of shares currently held. In practice, the issue of bonus shares even in that sense is fairly unusual. The next significant point of difference between shares in a cooperative and in an ordinary company is that the cooperative shareholder normally has only one vote, irrespective of the number of shares he may own; it is the man and not his money that counts. Finally, when the member of a cooperative society wishes to dispose of his shares, the society is able to buy them back from him; the reason being that shareholding in a cooperative is incidental to membership, and when the latter is terminated the former must be terminable also. The situation of an agricultural cooperative company is different in this last respect, since it is unable to buy back its own shares (if it is a company limited by shares; there are also a few agricultural cooperative companies which are limited by guarantee). Such a company will, however, wish to restrict membership to persons who have a trading interest, in the same way as a cooperative society. It is usual, therefore, for a cooperative company to issue no more than one share to each member. This share will be transferable when the member ceases to trade and, to encourage transfer, may carry the condition that the holder of it cannot exercise a vote if he ceases to have a valid trading contract with the cooperative. In such cases, where very few shares are issued, the bulk of the members' capital will have to be provided in the form of long-term or 'qualification' loans.

Shares in agricultural cooperative societies are repayable at the option of the society, not withdrawable at the option of the members (as is the case in most consumer cooperatives). This gives them some degree of permanence. However, there is a moral obligation on Boards to repay, as soon as possible, the shareholdings of members who have retired from farming, or who have for some other reason stopped trading with the society. If, for financial reasons, a society is unable to do this, it may consider giving such members a 'preferred' shareholder status. This question is discussed further in Chapter 6.

The characteristics of shares in a cooperative may at this point be summarized as follows:

- limited rate of interest
- no capital appreciation
- one vote only to the shareholder

These restrictions must clearly make members reluctant to subscribe for more shares than they are obliged to, particularly during a period of inflation.

Since, then, cooperatives must often work with an inadequate members' capital contribution in the form or shares, they have every reason for trying to build up reserves, the only other form of owned capital available to them, and one moreover that has the virtue of permanence. But here, too, there are problems. As noted earlier, any surplus remaining after the payment of a limited interest on shares will, if it is distributed to members, be receivable by them as a bonus on trading. If this surplus is, instead, retained by the cooperative and placed to reserve, there is no resulting increase in the value of the shares, as there would be if profit were to be placed to reserve in an ordinary company. On the contrary, unless reserves are distributed fairly soon after they are created, which, of course, is not the intention, the cooperative member must reckon that he will have no future claim on them. He must therefore always be from a personal point of view reluctant – to put the matter no more strongly – to allow the Board to make appropriations to reserves which it cannot justify in terms of current needs.

It is no mere affectation that in the preceding paragraph, as elsewhere in this book, the word 'surplus' has been used to describe the balance that is left in the hands of the cooperative after all its costs of administration have been met; in an ordinary company this would be described as its 'profit'. The difference is that, in a cooperative, this balance of funds represents the amount by which members have been overcharged (in the case of a cooperative supplying goods or services) or by which they have been underpaid (in the case of a cooperative selling their produce). In the final analysis, it is a question of policy for the cooperative concerned whether the initial price charged or paid to members is such as to leave a large balance, or surplus, or a small one. There is consequently some danger in comparing the surplus retained by a cooperative with the profit earned by a company. Inevitably, however, such comparisons are often attempted and, since most cooperatives adhere pretty closely to market prices, it is generally not unreasonable to make them.

INVESTMENT NEEDS AND OPPORTUNITIES

During the first half of the twentieth century, in fact until the Second World War, agricultural cooperatives were not and did not need to be very different institutions from what they had been in the nineteenth century. It is true that the substitution of mechanical for horse power transformed the economics of transport and distribution in the twenties and thirties, making it possible to serve wider areas from a reduced number of depots. But it was only after the war that other big changes began to take effect.

The first mainly post-war development to be noted is the growth in the average size of business serving or served by agriculture. In the pre-war era the normal type of business in this sector had been a family business, whether it was an agricultural merchant's private company supplying feedingstuffs, fertilizers and other principal agricultural inputs, or a butcher, greengrocer or corn dealer taking the farmer's produce. There were exceptions of course; the smaller firms pasteurizing and retailing milk, for instance, were already on the way out before the war, as the larger combines began to take them over. Post-war, many amalgamations took place in both the supply and marketing field, and the cooperatives too were strongly affected by this trend. The small, local cooperative has even now far from disappeared, but a large part of the total cooperative business is currently being done by cooperatives which have increased vastly in turnover, trading and membership.

Secondly, the modern agricultural cooperative is a far more capital-intensive organization than it needed to be in earlier times, when it was mainly a factoring concern engaged in distribution of supplies or collection of produce – as were most of its competitors – and only to a small extent in manufacturing or processing. The reasons for this development are the much higher standards of productivity which are now required, combined with the demand for more refined products in which the cost of the refinement is greater than that of the raw material. Modern agricultural cooperatives therefore require finance for their operations on a scale inconceivable a generation ago. But the members to whom they must look for this finance, the farmers, are engaged in what has itself fairly recently become a highly capital-intensive industry, even less able than most industries to spare funds for investment elsewhere.

Thirdly, monetary inflation has badly eroded many of the bases on which cooperation is founded. In particular cooperative orthodoxy still assumes that money can be remunerated by the payment of interest alone, whereas what the lenders of money now seek above all is an appreciation of their cpaital.

Fourthly, it becomes increasingly difficult to claim that agriculture is a primary industry; it is rather part of an industry which is converting one product into another, as the two sides of the balance sheet of the national farm clearly show.* It is necessary therefore for agriculture, while it remains independent, to become more closely integrated with the other sectors of the food industry. Cooperatives have an important part to play in this process, but need finance in order to undertake it.

* *Annual Review of Agriculture 1974*. The figures for 1972–3 are as follows: Final output £3053 million, net input £1393 million, gross product £1660 million.

THE COOPERATIVE RESPONSE

During the ten to fifteen years following the Second World War the growing financial pressures on agricultural cooperatives, though statistically evident in a declining proportion of owned capital (shares, reserves and undistributed surplus) to total capital employed in the business, did not give rise to overmuch concern. At the time membership of agricultural cooperatives was still growing, and although managers complained that it took a number of new members, coming in on the minimum shareholding, to compensate for the loss of one old one, whose shareholding might be at or near the maximum then permitted (£500), they were not too worried, since the purchase of shares by farmers had never been an important source of new capital. This was mainly provided by an almost universally adopted system of 'painless extraction', the procedure being to advise members that the end of year bonus due to them would be added to their share ledger account unless they informed the Society before a certain date they wanted it paid to them in cash. Most farmers were content to let the bonus lie, on the principle that they 'would not miss what they had never had', and the more successful cooperatives were able to retain a considerable part (often 75 per cent or more) of the bonus as new capital by this method, which had the added advantage of roughly equating the individual member's capital stake over a period of years with his trading throughput. A variant on the above arrangement was for the cooperative to declare part of its bonus to be payable in the form of shares, although an appeal to the Courts had to be made before it became established that under this procedure, as under the other, the additional shares were assessable as income of the producer concerned, and were a deductible expense in respect of the assessment of the society's own profits (Staffordshire Egg Producers Limited v Spencer, 1963 – TR 67 and 241).

During the late fifties and for much of the sixties the financial situation of agricultural cooperatives worsened, for a variety of reasons. In the first place there was a marked decline in the profitability of handling animal feedingstuffs and fertilizers, which constituted the largest part of their business, with the result that the high bonus rates paid in previous years had to be drastically

reduced and eventually, in many cases, disappeared altogether. Thus there was less bonus available to be converted into shares, even where members were prepared to leave it with the Society to accumulate, which to an increasing extent they were not, because they needed it for investment on their own holdings. Instead, demands for repayment of shares taken up in previous years increased, often in order to satisfy bank managers insisting on a reduction of their customers' overdrafts. Secondly, the capital requirements of the cooperatives themselves began to intensify, for reasons already examined. Development apart, many of them found themselves in the position that, in order even to retain existing business, they had to make drastic changes in their operating methods and, in brief, use less labour and more capital, all of which required substantial investments in buildings, transport and office machinery. A powerful incentive for them to make this reorganization was the competition they began to encounter during this period from low-cost buying groups, consisting of farmers who believed, rightly in many cases, that the on-costs of the 'traditional' cooperatives were excessive, and that substantial discounts would be obtainable if they were to club together and shop around on their own account.

Various steps were taken about this time to help the cooperatives to consolidate their position. In 1960 the Plunkett Foundation for Cooperative Studies published a *Survey of Capital and Credit in Agricultural Cooperative Societies in Great Britain* which, for the first time, exposed the financial problems of these organizations. The publication of this report (the *Morgan Report*) was followed by the introduction of a Private Member's Bill, which became law in the following year, raising the maximum shareholding in a cooperative from £500 to £1000. In England, which was the country chiefly affected by the financial problem, a committee was set up by the Agricultural Central Cooperative Association to examine and implement the report. This committee urged agricultural cooperatives to raise minimum shareholdings, reduce outstanding credit given to members, prune overhead costs and adopt more realistic reserve policies. Many agricultural cooperatives drastically overhauled their existing structures, and greatly improved their efficiency as a result.

THE MAXWELL STAMP, NORTON AND BARKER REPORTS

The *Morgan Report* dealt with finance from traditional sources, that is to say owned capital from members, working capital from members, banks and trade sources. In the early sixties the feeling began to gain ground that these sources were not enough. So far as short-term capital was concerned, it was generally accepted that this was best left to the banks. On the other hand, the banks did not, officially at any rate, lend for long-term purposes, and it was thought unreasonable to look to them to make good the gap in capital resources which, even on the most optimistic assumptions concerning share capital and reserve accumulation, appeared likely to arise in the future. Another cause for concern was the probable effect of entry into the Common Market, by then a subject of ardent political debate, on agricultural cooperatives in the United Kingdom. It was well known that within the six member states of the existing European Economic Community agricultural cooperation was in general much more strongly established, and it was suspected that the reason for this was the existence of powerful institutions specializing in credit for agriculture, from which the cooperatives in those countries were able to obtain valuable financial support. In the United Kingdom, by contrast, the only specialist financing organizations available to cooperatives (see Chapter 10) were very small and almost entirely lacking in 'owned' capital.

The investigation instituted by the national cooperative bodies resulted in a report (the *Maxwell Stamp Report*) of 1967. Its main recommendation was the setting up of an Agricultural Cooperative Finance Corporation to meet the serious gap the consultants saw developing in the provision of long-term capital, which the joint-stock banks would be unable to meet. The role of this organization would be to provide societies with long-term finance, to assist them with meeting demands for share repayments and possibly to act as a channel for the investment of farmers' surplus funds. The establishment of such a corporation was seen as providing an institution to parallel the agricultural credit banks that had long existed in the EEC, but it could not take place without government assistance, and the report therefore recommended that a substantial loan or grant should be made out of public funds in order to provide an initial reserve fund and shareholding for the Corporation, as well as recurring grants to cover its running costs in the early period.

The Maxwell Stamp conclusions were not accepted in their entirety by the Central Council for Agricultural and Horticultural Co-operation, which had been brought into existence between the time of commissioning the report and its submission to government, and which had been given the responsibility of advising the government on matters of this kind. However, while not commenting on the proposal to establish a specialist financial corporation, the Central Council did recognize the problem facing agricultural cooperatives by virtue of their dependence upon non-permanent capital and proposed (in the *Norton Report*) that, when both the cooperatives and their members had done all that could reasonably be expected of them, and all other normal sources of finance had also been used to the full, the government should be prepared to provide redeemable preference share capital at normal rates of interest, on a £ for £ basis. This proposal was rejected by the government, and there, for the time being, the matter rested. But not for long, for in 1972 the Barker Committee on Contract Farming came forward once more with the view that agricultural cooperatives in the UK were under a serious disadvantage, and that this situation 'can only be remedied by action which Her Majesty's Government would be entirely free to take'. The Committee accordingly proposed the establishment of an Agriculture and Food Development Authority, with investment functions which would include the provision on normal or special terms of risk and loan capital for cooperative groups, or for joint ventures between such bodies and private concerns.

On this occasion the Minister of Agriculture, while rejecting the proposal for an independent authority, invited the Central Council to advise him if cooperative marketing enterprises were being impeded by lack of investment capital, as the Committee on Contract Farming had indicated. This led to yet one more enquiry, carried out by a Working Party which the Council established, whose report was submitted to the Council in September 1974. Before examining this document it will be convenient to make a brief digression, so as to review the capital investment facilities from public funds which were already available to agricultural and horticultural cooperatives at the time the report was commissioned.

GRANTS FOR WORKS AND FACILITIES, AND OTHER INVESTMENT AIDS

The Central Council which administers these facilities will be described in Chapter 10. Capital grants are obtainable, on the Council's recommendation, under the Agricultural and Horticultural Cooperation Scheme (Statutory Instrument 1971 No. 419) by cooperatives meeting the Scheme's requirements – one of the most important being commitment of produce by the member to the cooperative – on a list of items broadly similar to those available to individual producers under the Horticultural Capital Grant Scheme (Statutory Instrument 1973 No. 1945) and the Farm Capital Grant Scheme (Statutory Instrument 1973 No. 1965). The horticultural grants cover a wider range of items than those for agriculture and are at a higher rate; the Council's latest annual report shows that during 1973–4 four-fifths of the total sum of £1·2 million approved for works and equipment was for horticultural purposes.

There is an added incentive to cooperatives to apply for official aid under this Scheme (or for aid under other schemes, such as that obtainable from the Department of Trade in respect of certain projects situated in development areas). This is that receipt of it qualifies the applicant for consideration under the guidance section of the European Agricultural Guidance and Guarantee Fund, in accordance with Article 11.1 (b) of Regulation EEC/17/64 ('adaptation and improvement of the marketing of agricultural products'). The importance of this source may be gauged from the fact that seven British agricultural cooperatives obtained aid from the Fund's 1973 allocation to the total amount of £426 000, equal to one-third of the total paid in 1973 to all of them under the 1971 Scheme. Aid from this fund is not available for horticultural marketing projects, which are considered (at the European level) no longer to require it.

A further form of aid for marketing cooperatives in respect of their investments is that available to them under a system of loan guarantees. During the period of negotiation for joining the

Common Market in 1970 the Central Council reviewed various forms of financing agricultural cooperation in the 'Six' and concluded that special attention should be given to such a system, which had been commended by the 1970 OECD survey *Capital and Finance in Agriculture*, was in operation in several of the Member States of the EEC and had been mentioned in various draft EEC regulations or directives as a type of financial assistance that would be in line with the common agricultural policy. After a number of investigations had been made, the government in 1972 authorized the Central Council to institute a pilot Scheme which after two years of successful operation was revised and enlarged into a new Scheme as from mid-1974. The purpose of the Scheme is to provide an additional collateral to cooperative marketing organizations which need bank finance but have little land or other assets to offer as security. The authority for giving such guarantees had been provided by Section 64 of the Agriculture Act 1967 (as extended by subsequent Orders). The guarantor bodies are the Agricultural Credit Corporation or the Agricultural Finance Federation, but the guarantees they give will only be underwritten by the government if they have had prior approval of the Central Council.

RECOMMENDATIONS OF THE WORKING PARTY ON INVESTMENT CAPITAL

Although the request made by the Minister referred specifically to marketing enterprises, the Working Party, in carrying it out, felt it necessary to collect information from a wide range of organizations, including multi-purpose cooperatives which, though primarily concerned with farm supplies, also had a significant marketing turnover. The investigation produced no evidence of any actual shortage of investment capital in marketing cooperatives, though it found there had been a decline in the proportion of total finance employed which was invested in long-term assets. An analysis made by the Industrial and Commercial Finance Corporation at the same time, but based on an earlier sample, showed that the risk capital available to marketing cooperatives was proportionately lower than that of non-cooperatives in the same line of business. These indications, and evidence obtained by the Working Party from interviews, seemed to show that, although marketing cooperatives were not being impeded by a lack of investment capital in their actual operations, other operations which would have been open to them were in some cases not being undertaken because the capital for investment would have been insufficient.

The Industrial and Commercial Finance Corporation survey also found that, as between the marketing group within the cooperatives and the non-cooperative group of private companies, there was a good deal of resemblance, in respect of both growth and profitability. It seemed therefore that differences in the financial structure of the two groups could not be attributed to any difference in the calibre of their management. This survey broke fresh ground, in that for the first time agricultural cooperatives were given an opportunity of comparing their own performance with that of businesses in the non-cooperative sector.

The recommendations of the Working Party were in three main areas. First, the report reviewed the scope for internal reconstruction and concluded that there was much to be done, in reclassifying members' shareholdings so as to distinguish inactive members from active members, and in encouraging the latter to contribute investment commensurate with their trade. (This had also been a recommendation of the earlier Working Party on Agricultural Cooperative Law.) Cooperatives were urged to go in for a policy of contracting with members, where they were not already doing so. Next, the report reviewed the sources of borrowed capital and suggested that cooperatives could make more use of special loans than they had been doing. It outlined ways in which cooperatives needing institutional loans could arrange to give the lending body a share in the profits and the risks of the enterprises being financed. Thirdly, the report reviewed the various sources of government assistance, making recommendations as to how the existing grant scheme and loan guarantee scheme could be improved. However, the main recommendations under this head concerned taxation. It was suggested that the primary need was to help cooperatives build up their owned capital in two ways. One was to defer the incidence of taxation on 'revolving funds' until members actually received the money due to them. The other was to lodge the corporation tax due on money placed to reserve in a separate fund, to be reinvested in preferred shares issued by agricultural cooperatives needing extra risk capital, and qualifying to receive it. At the time of writing, these recommendations were still being considered.

TAXATION OF AGRICULTURAL COOPERATIVES

Bonus, representing an addition to the price for goods produced or a reduction of the price charged on goods supplied, is deductable when computing corporation tax, either of a cooperative or of an ordinary company. Where share interest or loan interest are concerned, however, special provisions have been made for cooperatives. They allow such interest to be deducted when computing, for the purpose of corporation tax, what income has been earned. The relevant legislation, which is the Income and Corporation Taxes Act 1970, also allows companies which have been recognized as agricultural cooperatives to claim this dispensation.

It must be pointed out that the interest in question is none the less taxable, though in the hands of the member rather than in those of the cooperative.

In so far as surplus is divided among members, therefore, either as bonus or by way of interest, it is paid gross. Where, however, an allocation is made from the cooperative surplus to cooperative reserves, tax will be collected in the normal way. It has long been argued that such an allocation should be treated differently, in regard to tax, from allocations from company profits to company reserves, from which the individual shareholders stand to benefit, as well as the company corporately. An allocation to the reserves of a cooperative benefits the cooperative corporately, but to the individual members it represents a loss of their money in perpetuity.

The difficulty of building up adequate reserves through retention of surplus which is subject to corporation tax has led a number of groups engaged in farming cooperatives to consider the possibility of establishing organizations on a 'mutual' basis, by which the tax liability can be avoided, though only for as long as the reserve so created remains intact. The essence of mutuality is that a common fund is established. All the contributions to the common fund must be entitled to participation in the surplus and all the participators in the surplus must be contributors to the common fund. Following the Finance Bill of 1964, from which a proposal to tax the surpluses of mutual trading companies was eventually removed, it appeared that the principle of exempting 'mutuals' from taxation would thereafter have the support of both political parties. This provided a fresh stimulus for their development.

'Mutuals' may be either cooperative societies or cooperative-type companies and they may be concerned with any of the commercial activities carried out by farmers' cooperatives, although they have hitherto been created mainly among marketing organizations. In practice it will be necessary for the organization concerned to act as an agent for its members rather than as a principal. It is not possible here to go into all the details of the form of organization required, but only to mention that great care is necessary to ensure that the principles of mutuality are not breached at any point when applying them to a particular situation. It is also difficult in practice for a cooperative to convert from an ordinary to a mutual form of constitution, after it has become established.

NOTE: The reports mentioned in this chapter are the *Survey of Capital and Credit in Agricultural Cooperative Societies in Great Britain* by William Morgan, 1960, published by Basil Blackwell; *Financing Agricultural Cooperatives* by Maxwell Stamp Associates Limited, 1967; the *Norton Report* by a committee of the Central Council, 1968; the *Report of the Committee on Contract Farming*, published by HMSO in 1972 (Cmnd 5099); the *Report of the Working Party on Investment Capital for Agricultural Cooperatives*, published by the Central Council in 1974; a report comparing Agricultural and Horticultural Cooperatives with other Organizations, also dated 1974, obtainable from the Industrial and Commercial Finance Corporation Limited.

5. Agricultural cooperative structures

When, in 1974, the question what were the objectives of an agricultural cooperative was discussed with a number of directors and managers there was broad agreement on a definition as follows:

The object of an agricultural cooperative is to provide its members with a commercial service for the promotion of their economic interests, which will operate at the lowest cost consistent with the quality, maintenance and improvement of the service, and remain under producers' control.

The object of 'service to members' is consistent with the main principle of cooperative organization discussed in Chapter 2. The emphasis on its being a commerical service is noteworthy and indicates that in a country such as the UK, where agricultural cooperatives have only a minority share of the agricultural business, they must at all times be competitive. It is doubtful whether in countries where cooperatives are in a majority, even monopoly, position, this aspect would have been so much stressed. Finally, the reference to control by producers is a reflection of the unease felt by many farmers that their influence over their own industry may be on the decline. It is as well to recognize, however, that if there should be any conflict between the low-cost objective and the control objective, as there may well be, the former is likely to prevail. Cooperatives undertaking a new venture in which the chances of early success are problematical will therefore try to insist (and must insist, if they are in receipt of grant aid either under the UK scheme or under those existing or proposed for groups of producers in the EEC) on firm engagements by members, in order to safeguard their position. Although this question of member-cooperative agreements is of fundamental importance, it will be best to reserve it until some necessary distinctions have been made between cooperative developments in different parts of the agricultural industry.

STRATEGIC STUDIES

Starting from the premise that different sectors of the industry had totally different problems, and that the scope for cooperative development was likely therefore to vary greatly between one sector and another, the Central Council during the period of its first four-year plan (1969–73) initiated, in collaboration with a number of other sponsors (Farmers' Unions, Cooperative Associations, Marketing Boards, etc.), a series of 'strategic surveys' of the marketing outlook and development needed for many of the major agricultural and horticultural commodities. It was not found necessary to include milk, hops or wool in this list, since these commodities were already comprehensively covered by Marketing Board arrangements. Studies were, however, made and completed of the scope for cooperation in the marketing of potatoes (and, separately, seed potatoes), eggs, vegetables, cereals and pigs, and in the supply of requisites. All the studies were published and provided a basis first for discussion of the findings and, then, for working out the course of action to be pursued. Given the earlier lack of agreed objectives and coordinated action in the development of agricultural cooperation, this procedure represented an important advance towards determining what needed to be done and avoiding wasteful diversions of energy in doing it. The studies also confirmed the original opinion that no single method, or even objective, of cooperation would be found universally applicable to the different sectors of the industry.

It will be instructive to mention the findings of the various studies in order of their appearance, mentioning here only the conclusions relating specifically to organization, since marketing operations are dealt with in a later chapter. The first study to be undertaken was concerned with *Potatoes*. The conclusions were that the operation of grading stations by cooperatives was advantageous to their members, that the desirable capacity of a station lay between 10000–30000 tons within a fifteen mile radius, that some twenty-three grading stations were needed in England and Wales, and could be fairly reliably positioned, that preparation for market but not processing could be undertaken as a supplementary activity, that there was not much to be gained from a cooperative group operating more than one grading station, and that there was even less advantage in centralizing groups (as distinct from coordinating their activities). It is obvious that, if accepted, these conclusions were such as to provide a fairly precise guide to the sort of developments that needed to be undertaken and to rule out, almost from the start, any idea that 'second-tier' organizations (to quote a phrase much in vogue in recent years) were likely in this case to provide an appropriate solution. Although it appeared two years after the first, the study on the Scottish *Seed Potato* industry may conveniently be mentioned in the same context. The study recommended the development of grading and marketing groups (probably about eight of them, each handling about 800 acres of ware and 400 acres of seed) in the areas where they would be feasible, and the development of production and marketing companies where no such group was possible, or extending into the areas covered by the grading and marketing groups where these were unable to handle all the seed grown, and the establishment of an export company as a second-tier organization in which both producers' cooperatives and private firms would participate. The contrasts, as well as the similarities, between these two sets of recommendations hardly require comment.

The next study, which was concerned with *Eggs*, observed that the balance of power was swinging strongly in the direction of large associations of packhouses able to dispose of temporary surpluses, to market and promote national brands, to service national groups, etc. Since their costs depended to such an important extent on quality of eggs delivered, distance travelled, etc., packers must be tempted to create production facilities for themselves rather than depend on existing facilities which were badly sited or otherwise inadequate. Such a solution would clearly present a cooperative with an acute 'ethical dilemma', which might be solved by some form of joint enterprise between the cooperative and its members, though this would involve the members concerned in losing control over their production businesses and ending up as shareholders. The consultants, in the words of their report, 'recognize the obvious fact that such a solution would be unwelcome to many producers, but ask the question – if our analysis is right, and if a man is determined to remain associated with the industry, what other solution is there?'

A study of producers' *Vegetable* marketing organizations prepared in 1972 dealt with a sector of the industry in which there had been revolutionary changes. On the production side there had been a transfer of acreage from traditional market-garden areas to arable areas, where some crops formerly grown intensively were now being produced on a field scale. On the marketing side, the power of independent wholesaling firms had declined and a new type of unified supply organization had emerged whose capacity matched that of the big retail organizations which dominated the distribution field. Between the producer and the market, the study suggested, there would be a growing number of packhouses concerned more with packing than with grading, whose operation would in fact be economic only if a high proportion of the produce handled was of packable quality. In a great many cases it would be more logical for the packing operation to form part of the unified supply organization (which would also be dealing in imported produce) than part of the production organization; in other words, the report argued, produce would tend to change hands before entering rather than after leaving the packhouse, in which case cooperatives in vegetables would tend to find a role mainly as supplying organizations, rather than as organizations fully involved in packing and marketing operations. This conclusion of the vegetable study was clearly as challenging to existing concepts of the future scope for cooperative development as, in a different way, the study on eggs had been.

Following the report on vegetables, two studies were made of the potential for cooperative development in the marketing of *Grain*. The conclusions of these studies, though made by different consultants, were along the same lines as those reached in respect of potatoes. An increase in the volume of grain being handled by cooperatives was foreseen, but it was thought that the number of organizations involved would tend to decrease. There was no reason to suppose that a trading federal, if formed, would be able to extract any higher prices from the grain processing industries than a large multi-purpose society could obtain. Nor was there any case of creating a further export organization. The conclusion that a trading federal should not be formed was supported by the second study, which thought, however, that arguments could be found for forming a servicing company with responsibility for central organization and development of grain groups. The conclusions of these studies were important because, at the time they were produced, firm convictions had been expressed that the formation of a 'second-tier' organization was essential to cooperative development in the cereals sector.

The last of the commodity sectors to be made the subject of a strategic study was one carried out in 1973 on *Pigs*. Here, the evidence pointed to the need for a much closer relationship than in the past between the feed manufacturer, the pig producer and the processor, with the three parties becoming increasingly interdependent. However, such a relationship could not be made to work unless producers became better organized. Secondly, the report concluded that the appropriate point for transfer of ownership from producer to market supplier was after the pig had passed through the slaughterhouse and had been converted into a carcass or parts of a carcass, rather than while it was still alive. This conclusion, like the one relating to vegetables mentioned earlier, was in marked contrast with traditional and officially held views on the function of producers in relation to their markets.

Finally, a study was carried out in 1972 of *Requisite Cooperatives* with the object of forecasting their future role in the marketing of animal feeds, having regard to the changes occurring in the feed industry, including those likely to arise from entry into the EEC. The report dealt with the existing and possible future share of agricultural cooperatives in the manufacturing and marketing of animal feed, the case for manufacturing and whether this should be considered on an individual society or on a group basis, the type of mill to be constructed, and the relationship which the society needed to develop with its members to make such an operation worth while.

CONCLUSIONS TO BE DRAWN FROM THE STUDIES

The intention of the Central Council in promoting these strategic surveys and inviting national organizations concerned to act as co-sponsors for them was to obtain an agreed point of reference from which discussions of the policies to be adopted in the various fields covered by the surveys could start. In a sense it did not matter too much whether the conclusions drawn by the independant consultants appointed were acceptable or not, the point was that they were as good as could be got at the time with the knowledge available and were free from any suspicion of having been advanced with a particular political end in view. On the whole it can be said that the surveys have well fulfilled their initial aim. In some cases the production of the study has been followed by a continuing dialogue between the sponsors on the policies to be pursued but, even where this has not happened, it has helped to clarify objectives and remove misunderstandings. There is in consequence less danger of contradictory or unnecessarily competitive approaches to problems being adopted than there was before the studies appeared.

A striking fact about the different surveys is the variety of organizational solutions they recommend. However, there were common factors as well, though not so immediately apparent. In the first place the surveys shed new light on the old question whether future development should be based more on single or on multi-purpose organizations. The argument hitherto had been that the growth of specialization in production, and accompanying higher productivity, called for correspondingly specialized marketing institutions; moreover a specialist organization would find it easier to insist on trading commitments from its members, through being able to identify more closely with them. On the other hand it had also been argued that multi-purpose cooperatives, particularly the larger requisite societies, with more reserves, better access to external finance and less dependence on external booms and slumps, would be better situated to undertake the risks involved in new marketing enterprises. The statistics do indeed show that the marketing element in the turnover of these societies has grown at the same rate as their requisite trading, having remained constant at 12 per cent of the combined marketing/

requisite total during the three years 1970–2. Indeed, these societies did better during this period than the specialist marketing societies (whose returns were badly affected by the lower prices for dairy produce and eggs) though less well than the specialist marketing companies.

Marketing turnover £000 (Source: Plunkett Foundation)

	1970	1972	% change
Requisite cooperative societies	23029	28875	+25
Marketing cooperative societies	97771	92433	−5
Marketing cooperative companies	39600	51303	+30

There are, however, certain inhibiting factors for the requisite society which may consider widening the sphere of its operations to undertake more marketing. The first is that the proposed marketing activity, in which the objective of the cooperative is to obtain the best price for a member's produce, may well conflict with its supply activity, in which its objective is to buy as cheaply as possible on their behalf. A case in point is where a cooperative has to sell a member's grain, but also needs to buy grain from farmers for conversion into feedingstuffs. A further question which the requisite cooperative may ask itself is whether it would be justified in using capital derived from all of its members for a risk-bearing marketing project which will benefit only a certain number of them. It may take the view that it would be right to give this new enterprise some encouragement, but that a fair proportion of the cost involved should be borne by the members who will make use of the facility once it is in operation. This must lead it to consider whether some form of subsidiary could be created, under separate specialist management, and with a board who could give its problems the detailed attention which the main board of the cooperative cannot possibly do. It has been difficult to find a formula which would enable the parent cooperative to exercise the degree of control over such a subsidiary that it would like, without the subsidiary losing the degree of independence that it must possess if it is to be recognized as having a properly cooperative constitution. Recently a formula has been proposed which will go some of the way towards resolving this difficulty; as mentioned in Chapter 3, it is that local cooperative bodies should be set up with limited objectives which would not be alterable without the consent of all the members, including the 'parent' cooperative. This would at least remove any risk of a subsidiary later developing in such a way as to bring it into conflict with the body that had set it up, which has often been cited as one of the chief problems. It remains to be seen whether this formula will satisfy the multi-purpose cooperatives, even though many of them are beginning to feel that more devolution of their multifarious interests ought to be made, in the interests of general business efficiency.

A second point of interest about the studies is their general lukewarmness towards second-tier organizations. This was at first rather surprising. The belief had grown up, based on experience with apples and eggs, that first-tier cooperative organizations concerned with the functions of procurement, grading and packing, limited in size as they had to be both for geographical reasons and also by the desire not to become so big as to weaken contacts with their members, could advantageously surrender their marketing functions, in which size was important, to a federal or second-tier cooperative. On the evidence of these studies it would seem that the need for second-tier producer marketing organizations is to a considerable extent dependent upon the degree of centralization of the processing and distributing firms which they serve. It should be borne in mind, however, that the authors of the studies were primarily concerned with providing guide-lines for the development of base structures rather than with what might follow once that base had been formed.

A third common factor which runs through all the studies is an insistence on legal understandings between the members and the cooperative to which they belong, preferably in the form of written agreements or contracts, although there may be a tacit understanding that the purpose of these is to define the obligation rather than to facilitate its legal enforcement. (Such contracts between a cooperative and its members are quite distinct from those between a cooperative and its market partners, which will be discussed in Chapter 11.) Contract arrangements have received some publicity, since the Central Council has made them a condition of grant. But that is, or ought to be, purely incidental, for in a business that was truly cooperative in spirit the loyalty of members should be taken for granted. In practice it cannot be. The reasons for this lie deep in legal history. If it had been necessary for British cooperatives to be formed on the basis of unlimited liability of members, as was normally the case on the continent, a member-cooperative contractual situation would have automatically ensued, since the risk of failure due to members' lack of support would have been inadmissible. Instead of this, the liability of members has been limited to the amount (usually an insignificant amount) of the shares held by them. Since the loss of this provides no penalty for disloyalty, the cooperative has to be protected in other ways. Hence the importance of binding agreements, which can give it that protection.

A fourth conclusion which emerges from the studies is concerned with the point of transfer of produce from the producers' organization to its market partner. It is logical that the area of operation of a marketing cooperative should be that extending from the production point to the transfer point; agricultural cooperative interests have always argued for recognition of this fact, whereas the policy of government has tended to give no encouragement for cooperative activities extending beyond activities 'normally associated' with farming and growing, i.e. activities in theory capable of being carried on, though less efficiently, by farmers and growers individually. The legal framework of this policy is given in Chapter 8. Such a formula has always seemed to cooperatives to be too restrictive, as it takes no account of the fact that 'raw' agricultural produce is less and less acceptable to the market, and that the operations for converting it into a marketable state are more and more beyond the capacity of an individual farmer (see Chapter 1). The essence of the matter would seem to be whether the conversion operation is wholly or mainly concerned with the farm product; if it is, the distinction between an advanced or a simple operation does not seem to be too important.

COOPERATIVE PLANNING

The planning of agricultural and horticultural cooperation has mainly been carried out at the local level. Each Board of Directors and manager of each cooperative decides where, how and for what purpose it will develop, using such resources as it may have available. In doing so it may or may not consider the interests of other cooperatives already established in the area, together with any advice it may have received from its central association – presuming that the central association has been consulted. Such limitations on its freedom of action cannot be said to be very significant. In recent years other elements have entered into planning, such as the availability of government grants, combined with the greater use now being made of management consultants. In 1969, in execution of powers given to it under the Agriculture Act 1967 to coordinate cooperation in agriculture and horticulture, the Central Council set up a cooperative planning unit, which has sought and obtained the agreement of the cooperative Central Associations and Farmers' Unions on planning objectives, and has been able to stimulate a fair amount of collaboration between cooperatives in various fields. The strategic studies mentioned earlier in this chapter are among a number of coordinating activities that it has initiated. In 1974 a further extension of the Central Council's responsibilities occurred which, as will be explained in Chapter 10, laid additional emphasis on its planning function.

An obvious difference between the agricultural cooperative structure of the United Kingdom and that of most other countries is the relative weakness in the former of federal organizations. This is both an effect and a cause of the decentralization of planning already mentioned. When, as noted in Chapter 2, the Agricultural Wholesale Society collapsed in the 1920s, taking the Agricultural Organisation Society down with it, the English cooperative societies were left without any focal point, either commercial or political; the Cooperative Wholesale Society, being primarily concerned with consumer cooperatives, could not fulfil the first role, while the National Farmers Union, being primarily concerned with the interests of producers, could not perform the second. Consequently individual societies, during this critical period of their history, tended to turn inwards upon themselves, seeking through the growth of their own organizations the benefits of scale which, in other countries, have been obtained by concentration of individual cooperative interests in federal organizations.

Another vital factor influencing the form of development of agricultural cooperatives in the United Kingdom has been the absence there of any central financing institution. Where such institutions exist – as they do in virtually every country in Western Europe – endowed with important but, nevertheless, limited resources, which they aim to use profitably, to their own best advantage and to that of their client cooperatives, they must in practice do a great deal to guide the policies of individual cooperatives by making available or withholding the means of investment. It is true that the financial institutions in question do not claim to have this planning function, and might be reluctant to admit that they operated in this way. It would, however, be altogether surprising if they did not, and it is not difficult to find evidence of the exercise of this function in practice.

AGRICULTURAL COOPERATIVES AND MARKETING BOARDS

In Western Europe, and, indeed, throughout the world, agriculture is habitually accorded a special treatment under national policies such as is received by no other industry. In part, this is because it is concerned with one of the necessities of life. An equally important reason, however, is that when, following the industrial revolution, most other production industries at the same time as they adopted new technical processes also coalesced into a relatively small number of units, the average production unit in agriculture, though it increased in size, still remained centred, for the most part, round the family farm. Governments, being disposed to accept this as a necessary and even desirable state of affairs, have had to concern themselves with the resulting problem of how to enable agricultural producers to make good, in their trading capacities, the negotiating weaknesses from which they would be bound to suffer if they dealt as individual buyers or sellers. ('Buying retail and selling wholesale', was the phrase in use to describe this situation in Britain in the 1930s.) Hence arose various policies for the promotion of marketing boards, of cooperatives, or of producers' groups (in the EEC sense), all of which are variations on the same theme of coordinating and strengthening the marketing efforts of producers, though emphasizing different aspects of the process. The producers' group variation will be examined in Chapter 12. Here it is necessary to say something about cooperatives in relation to marketing boards, both of them being institutions with a British parentage.

The fundamental difference is, of course, that cooperatives are voluntary organizations, and except in the rare (and in the United Kingdom non-existent) cases where they have achieved a monopoly, they have continually to justify themselves on the basis of providing a service that the farmer cannot afford to do without. Marketing boards, on the other hand, are the sole buyers of the product, and the lesson has sometimes been drawn from the experience of the British Egg Marketing Board that a board monopoly must be complete in order for it to be effective. When the prospectus for British entry into the Common Market, the White Paper *The United Kingdom and the European Communities* (Cmnd 4715) stated that the UK marketing boards 'are expected to continue their essential marketing function', this phrase was widely construed as meaning that the monopoly powers of the the boards, the essential element distinguishing them from voluntary cooperatives, were not considered to be at serious risk.

Others, however, have taken the view that in practice, and particularly as imports of produce from other parts of the Community begin to have a greater effect, the differences between marketing boards and cooperatives will tend to become less marked. On this view the marketing boards, like the cooperatives, will rely not on the statutory controls they can exercise, but rather on the added value given to the product (e.g. in the way of collecting, grading or packing it) to retain the loyalty of their suppliers and the custom of their clients. One may judge from the generally favourable light in which marketing boards are regarded that their fairly small costs of administration are considered to be more than offset by the various services they provide. In certain cases, for example where the British Wool Marketing Board is concerned, one could say that these services have become so indispensable both to producers and to buyers of wool that, if at any time the Board were forced to wind up as a statutory organization, it would immediately be resurrected as a voluntary one.

Thus, considering the whole situation from the point of view of efficient marketing (it being recognized that marketing boards may also have supply-management functions, which raise entirely different issues), one may look forward to the probability that the marketing boards will rely less and less on statutory powers, and more and more on marketing services which are closely adjusted to the needs of the producers and the market; in other words they will become increasingly akin to marketing cooperatives in their orientation if not in their constitution. Although present developments could not have been foreseen at the time of the passing of the first Agricultural Marketing Act in 1931, it is significant that the preamble to that Act contains a reference to cooperatives and that the powers attributed to the boards under their various schemes all include powers to encourage cooperation among producers; the drafters of the Act seem to have been far sighted enough to realize that the statutory bodies which they were setting up must, eventually, establish themselves as marketing organizations based on the consent of sellers and buyers.

NOTE: The various Strategic Studies mentioned in this chapter were published by or are obtainable from the Central Council.

6. Members, Board and Management

A main theme of this chapter, as of Chapters 3 and 4, will be the differences between a cooperative and an ordinary company. The internal relationships within a cooperative, which is an association of persons, are in some cases like but in other cases quite unlike the internal relationships within a company, which is an association of capital.

MEMBERS

The members of an agricultural cooperative possess the double status of customers who are primarily interested in fulfilling their own needs for goods of the best quality at the lowest price (or, if they are selling their own produce, in obtaining the highest price for it) and owners of the business whose interest should be centred on its being run efficiently and profitably. In theory these two separate interests will be brought into harmony. In practice the member's interest in the business as a co-owner is not always equal to his self-interest as a customer. This can lead to problems for the cooperative manager since, if members are dissatisfied as customers with management decisions, they have certain opportunities as shareholders for changing them by putting pressure on the manager, who may find himself in the anomalous position of having to defend an organization in which he is merely an employee against attacks from those who own it. It is for this reason, presumably, that the law has stepped in (in many countries, but not in the United Kingdom) to ensure that certain precautions are taken, such as the creation of a statutory reserve out of a part of any annual trading surplus, so that the Society will not be weakened as a result of unreasonable pressures from its own members.

No parallel to this situation exists in non-cooperative businesses where, although there is the same basic conflict between shareholders and customers, they are for the most part two different sets of people. Most cooperative managers would agree that the management of a cooperative is made more difficult by this shareholder/customer relationship, though there are other compensating factors. An additional conflict of interests arises where the members are not only shareholders and customers but employees as well. This is the situation in a number of consumer cooperatives, but it is not one which the managers of agricultural cooperatives have to confront. In a number of agricultural cooperatives, as was mentioned earlier in Chapter 4, a proportion of the members have ceased to be customers but have continued to be shareholders, for the reason that they have not asked for their shares to be refunded or, if they have asked, it has not been possible to repay them. Such persons are in an unenviable position, since they are primarily investors, in an organization where the object of making profits mainly for the payment of interest on investments is expressly excluded. It has been suggested that, to get rid of this anomaly, a special status of shareholding should be created which would enable them to earn at least a higher rate of interest than the still active members, who obtain benefit from the cooperative's activities through their trading activities. A preferred shareholder would, however, have to surrender his right to vote on the ordinary business of the cooperative.

An agricultural cooperative society will normally include in its rules the right to repay a member's shares in various circumstances, one of which is his ceasing to do business with the Society. It is in the interest of both parties that it should do so, of the member because the shareholding without the business activity is an unprofitable investment, of the Society because it will not normally wish to have inactive members exercising an influence on the policies of an organization which is being run in the interest of trading members. The cooperative which is incorporated under the Companies Act is in a difficulty in this respect, since it can only transfer shares and not repay them. But it is not impossible to get round this difficulty, and a method of doing so has been mentioned in Chapter 4.

Finally, it follows from the definition of a cooperative society as an association of persons that each of its members has only one vote, irrespective of the amount of his shareholding. However, this rule of strict democratic equality has been modified to the extent of allowing members who are under contract to the co-operative for the marketing of their produce to have additional votes in proportion to throughput, so long as no member has more than one-tenth of all the votes that can be cast, and only one vote on any motion to amend the rules. In the case of a federal, the distribution of votes is a matter for the member organizations themselves, but in principle should be related to the number of individual beneficiaries from the federal's services within their organizations.

BOARD OF DIRECTORS

The Industrial and Provident Societies Act recognizes the existence of a committee, defined in the Act as 'the committee of management or other directing body of the society'. Every cooperative society must have such a committee, which in agricultural societies is normally styled a Board of Directors, in order to make it clear that their task is direction, as distinct from management. Every committee or board member is held liable for any offence committed by the Society of which he is not proved to have been ignorant, or to have attempted to prevent, unless responsibility for that matter has been given to a particular officer under the Society's rules. An 'officer' may be a treasurer, secretary, member of the committee, manager or servant of the society other than a servant appointed by the committee. Unlike a company director, who does not need to be a shareholder unless the Company's articles make this a qualification, a director of a cooperative society registered under the Industrial and Provident Societies Act must be a member, and therefore a shareholder. This means that when a cooperative society seeks, in accordance with modern practice, to reinforce its lay board with a paid managing or executive director, brought in from outside, it must first arrange for the individual concerned to take up membership of the Society in order that he may be eligible for this post. In passing it may be noted that the appointment of managing director is sometimes preferred to that of general manager, partly for prestige reasons and partly as a precaution against the Board taking decisions, which might have an adverse effect on the Society's future, in the absence of its chief executive officer. These reasons are not altogether convincing. The general manager or managing director of a cooperative society quite often doubles his post with that of secretary, in view of the fact that certain of the secretary's official duties, such as the repayment of shares on the order of the Board, have important implications on the society's overall policy, and ought not to be divorced from it.

While company directors in private companies often owe their appointments to the number of shares they control, in public companies they usually have only a small shareholding. In the latter case it is their salaries and other emoluments, rather than their dividends, which constitute their main sources of profit from the company, with the result that management and ownership become divorced. In an agricultural cooperative society the situation is different again. Few if any of the board members could be described as professional directors; their professional skill and knowledge belongs to quite another sphere, namely that of agricultural production, and their business sense often has to be acquired after rather than before they take their places on the Board. This has both advantages and disadvantages as far as the manager is concerned. The disadvantages are that a board consisting of practical farmers can hardly help looking at policies (e.g. pricing policies) as these reflect on their own farming businesses, rather than on the needs of the cooperative (e.g. to trade with adequate margins). Or they may be persons of limited outlook, unable to raise themselves to the level demanded of them. This is much less common nowadays than formerly, when one would hear of boards unwilling to pay salaries adequate to the responsibilities of managerial posts, mainly because the managers would then be earning more than themselves.

However, the situation has advantages as well as disadvantages. One is that the competent manager has at least no danger of professional rivals on the board and, provided he is not afraid of responsibility, this places him in a very strong position. In practice one finds that many Boards of Directors have a deep respect for their manager, and will defer to his judgement to an extent which may even be excessive. The wise manager is the one who does not profit from this situation to become a dictator but, on the contrary, takes the greatest pains to ensure that the directors on his board are fully associated, both in heart and mind, with the decisions that are taken in their name, and will defend them, however unpopular they may be, with the rank and file of the membership. The relationship between Board and Management is a subtle one. It is understandable that a 'good' board, consisting of capable farmers with an appreciation of business problems, is more likely to appoint a competent manager than a board consisting of men with a limited outlook. What is more unexpected is that a capable manager is often able, in the course of time, to attract a good board. This is not only because he can do a good deal himself to bring out the best in the Board he has inherited but also because farmers who have business competence are more likely to allow their names to go forward for election to the Board if they see that the affairs of the society are efficiently administered and feel that the talents they possess would not be wasted.

With a lay Board of Directors and professional management it is obviously necessary that there should be a clear division of functions. This is sometimes expressed in the proposition that the directors should be responsible for deciding policy and the manager for carrying it out. The real-life situation cannot be described so simply. Policy-making is hardly ever a single, dramatic gesture. It is more often the slow sorting out of possibilities into probabilities and probabilities into actualities, in which Board and Management are intimately associated as a team.

Nevertheless, some broad definitions are desirable. These were well stated in the Gaitskell Commission report of 1958 (which dealt with consumer cooperatives), as follows. The proper responsibility of the Board embraced:

first, the selection of the chief officials, and the determination of their salaries and conditions of work. This must always represent both the final and decisive power, and the most exacting duty, of the elected Board. In the existence of this power lies the ultimate guarantee of democractic control; on how judiciously it is exercised, the efficiency of the society will depend. We do not, however, believe that the Board should appoint officials below the top level – certainly not, as happens even in some large societies today, the branch managers; it is a function of executive management, once appointed, to select the subordinate officials.

Secondly, the Board should lay down the general policy on prices, dividend and reserve allocations. Thirdly, it should decide the rate of interest to be paid on members' capital. Fourthly, it must sanction all capital development for the coming year or years. Fifthly, it must approve the capital budget showing in detail how the plan is to be financed. Sixthly, it must receive from the chief official, and carefully scrutinise, regular trade and statistical reports designed to give the maximum information and the most scientific check on efficiency; it is in carrying out this regular scrutiny that the Board fulfils its main responsibility to the members, and acts as their trustee for the affairs of the society. Lastly, it must provide a continuous liaison with the membership. . . .

This summary can be criticised in that it does not spell out clearly enough the responsibility of the Board for ensuring that the society has clearly formulated objectives, which are regularly examined and revised. In general, however, it is a satisfactory statement of a cooperative board's functions.

MANAGEMENT

The passage just quoted continues as follows:

Once these duties are accepted as being the proper responsibility of the Board, societies must then delineate as clearly as possible what is the proper function of management, so that the division is fully understood by both Board and officials. We have been surprised to discover in how few societies is there any clear definition of the responsibilities respectively of the Board and its chief officials; in the rare cases where the duties of the latter are laid down in the rules, these are often completely out of date. The consequent ambiguity inhibits rapid and decisive action by the officials, who are uncertain as to what does and what does not lie within their competence.

The position of the manager in many agricultural cooperatives is not so very different today, according to a recent Wye College report, which finds that:

farmer directors tend to acknowledge their obligations as Board members in general terms, rather than specific supervisory duties. A significant minority proved unfamiliar with the procedures by which their society's operations were regulated, especially in the reviewing of past performance. . . . Role definition has been left particularly obscure in the case of the cooperative manager. Board supervision is most often asserted by limiting his discretionary powers. Far more rarely is real control over operations achieved, by requiring from him systematic records to demonstrate how effective past performance has proved in the market context.

Although good management is a key factor in the success of cooperatives and although it is generally claimed that the management of cooperatives presents problems of a kind different from those encounted in the management of other, non-cooperative businesses, no serious attempt has been made so far as the agricultural sector is concerned to define what these differences are. The crux of the matter seems to lie in the fact that, as a cooperative is not formed primarily for the purpose of making a profit on the capital invested in it, the main yardstick for assessing the efficiency of business operation, namely return on the capital employed, cannot be wholly applied to it. It is true that the management may be given a general directive not to provide a service unless it is profitable, or can be made so, but the interpretation of this in practice will not be easy. The cooperative manager will constantly be confronted with the decision whether a facility, say a depot which is conveniently sited for members but inconveniently for the cooperative, should be retained or closed down, whether the supply of commodities, which are essential to his members but yield a very low margin to the cooperative, should be expanded or discontinued, whether the member whose produce falls short of the standards required if the cooperative is to maintain its profitability norm should be tolerated or invited to take it elsewhere. In all such matters, the straightforward pursuit of profit has to be qualified by a number of other non-profit considerations, usually, though imprecisely, described by the word 'service'. Of course, these non-profit considerations also apply to company policies as well, but less strongly than they apply to cooperatives, where the member's interest as a shareholder takes second place to his interest as an individual entrepreneur.

The difficulty in defining management objectives has as its corollary a difficulty in determining management rewards in a cooperative. How are cooperative managers to be paid? Should the top man or men receive a salary which is predetermined in amount, being more or less related to the size of the business? Or should it contain an element – which will in practice need to be a major element – varying with the success or otherwise of the cooperative's affairs? In that case, how is success defined in relation to the cooperative's trading? Does it simply mean the volume of its business or turnover? Certain cooperatives thought so, and introduced an appropriate formula into the calculation of their manager's salary; as the turnover of the society increased, so did his remuneration. Some of the societies which used this formula became very big indeed, and their managers well known for their empire-building propensities! Alternatively should salaries be dependent on profits? This too can be unsatisfactory, for while there may be good cause, in some situations, to aim for a substantial margin of profit and a high rate of bonus, there will be other times when the opposite policy will be a correct one. In deciding what their policy will be, cooperatives have to consider very carefully what incentives they will need to provide to encourage the management to pursue it.

In this connection, there is one other point to be mentioned. As any owner of shares in public companies will be aware, their boards have long since appreciated that the present-day manager is not to be enticed by salary alone, but requires to be given other inducements which will enable him to accumulate capital in a form less vulnerable to inflation. Hence a variety of company schemes for their senior staff, the effect of which is to link their remuneration with the equity of the company, which as a result of their efforts is expected to increase in value. With a cooperative, the value of whose shares stands resolutely at par, such schemes cannot be operated. This must give them a disadvantage in the competitive field of recruitment, which has to be made good in other ways.

Most of the problems that the agricultural cooperative manager has to contend with are similar to those faced by a manager of any other business, and this is not the place to describe them. The main differences are, as already mentioned, the limitations imposed on him by the special situation of cooperatives in connection with the raising of capital and credit, his relationship with a non-professional board, and the attitude to management of customers who, in another capacity, are the owners of the business. The third of these factors gives management of cooperatives a dimension not found elsewhere, since it imposes on the manager, and indeed on staff at all levels, the necessity of continuously learning from the members what it is that their individual production businesses require, and of explaining to them in turn what their cooperative trading business is able to offer. This means that good processes of communication, which are important in any trading concern, are to an agricultural cooperative absolutely vital.

POLICY-MAKING IN AGRICULTURAL COOPERATIVES

Many farmers, while accepting that the object of their agricultural cooperative is formally as stated at the beginning of Chapter 5, doubt whether their interests are in reality its only or even its most important object. Furthermore, they question whether, even if the shareholders of these cooperatives consist solely of farmers, these farmers can in any real sense be said still to control them. They would maintain that when a cooperative grows beyond a certain size it becomes the tool of those who work for it rather than of those who employ them. This statement is, of course, emphatically denied by cooperative employees. However, neither the original claim nor its outright denial are completely credible.

It is inevitable, and entirely legitimate, that other interests than those of the owners of the equity should be taken into account in the operation of any business enterprise. Summing up at the end of a national conference of British Junior Chambers of Commerce held under the patronage of HRH the Duke of Edinburgh in 1962, a Mr Goyder said to the participants:

> . . . you have said in your report and discussion that we must start with the proposition that Management is responsible not only to the shareholders of industrial companies but also to the workers in them. You've gone further and said that from now on Management also has a responsibility to the consumers and community. However this fourfold responsibility of management is to be reconciled, it represents a change in the doctrine of management. It cuts across the assumptions of our

present industrial structure, both in nationalized and private industry. Your doctrine denies that a business belongs only to its shareholders. With this I warmly agree.

The same doctrine applies to agricultural cooperatives. The cooperative which constitutes an essential business service to its members is an equally essential means of livelihood to its employees and source of supply to its customers. Those who lend the cooperative money have a right to ask that their financial interest should be protected, and the community which harbours it that the environmental interest should not be overlooked. If it could be generally recognized that the cooperative in action must be a balance rather than a conflict of interests, this would come nearer to the reality of the situation. It is true however that a correct balance can only be achieved if the various forces which are involved all play their proper part. To judge from the poor attendances at general meetings and the almost universal system of appointing to vacancies on the Board only those whom the existing directors have agreed to coopt, the influence of the farmer members of many agricultural cooperatives falls far short of that to which they are entitled, and ought to be using.

EDUCATION AND TRAINING

The failure of the democratic machinery to work as it should is not the cause but rather the consequence of a widespread malaise affecting British agricultural cooperatives. This has been diagnosed as a fundamental lack of understanding on the part of British farmers as to what are the functions of a cooperative, what it can and what it cannot be expected to achieve. There are many factors contributing to this, some of which have been mentioned in earlier chapters. They include the legal system, which limited a member's stake to the size of his shareholding, the disrepute into which agricultural cooperation fell in the period between the wars, which agricultural leaders did little to correct, the easy alternative offered by statutory marketing boards, the disincentive to initiatives in marketing resulting from a price guarantee system, the neglect of past governments, and the lack of interest on the part of agricultural educators and trainees. Most potent of all, perhaps, was the fear of farmers that agricultural cooperation was in some way 'political', a fear which was without any basis, but was played on to the full by the commercial rivals of cooperation, using arguments which the farmer never bothered to verify.

It has become easier to spell out the causes for the comparatively slow rate of agricultural cooperative development in the United Kingdom now that serious efforts are in hand to deal with them, and there can no longer be any suspicion of finding excuses to cover up inactivity. The results of these efforts will be described in later chapters. As will have been seen from earlier ones, the impetus to make them has come from various sources – the prime mover in most cases being entry into the Common Market, which has brought about a re-orientation of many established patterns of thought. So far as government policy is concerned, the setting up of a Central Council for Agricultural and Horticultural Co-operation in 1967 was a decisive step in recognizing the problem and in attempting to find a solution for it. The new policy was backed with grants, the intention of which is to stimulate the main initiative for self-help, which can only come from farmers themselves. In the ultimate historical analysis, the effect of this government policy of encouraging agricultural cooperation will have to be judged in terms of the extent to which farmers own attitudes towards cooperation have changed as a result of it.

Grants towards production and marketing projects will enable schemes to be begun (or enlarged) which will be of help to the farmers who participate in them, or who may join them later. These schemes if successful will encourage others to follow their example. So there is a cumulative effect here, even if it acts somewhat slowly. Another line of action, also slow to produce results, but ultimately more cost-effective, is to promote the idea of cooperation through training and educational programmes; if there is to be a wholehearted adoption of cooperative techniques, they must be learnt at an early stage in an individual's experience. The Central Council has therefore given a high priority to educational schemes, which operate at the following levels:

- in most agricultural institutes, case studies based on the experiences of cooperative production or marketing groups now form part of the educational syllabus
- at Universities and Agricultural Colleges, cooperative studies

are normally introduced into the teaching of agricultural economics or agricultural marketing
- special courses are held for groups such as bank managers, advisory service staff etc
- farmers are encouraged to get to know about cooperation by farm walks, films, discussion groups etc
- seminars are arranged on a variety of subjects; the 'closed' seminar, in which the Board and senior staff of a cooperative take time off to discuss the growth and future of their organization, has proved particularly popular
- specialist courses are arranged for managers

The Central Council, which has taken on itself the responsibility for organizing these courses, is fortunate to have the assistance of the Plunkett Foundation. This foundation, established by Horace Plunkett in 1919, has a tradition of activity in the field of agricultural cooperative education and research on which there is a good possibility of building for the future.

NOTE: The *Cooperative Independent Commission Report* (1958–Cooperative Union Ltd), though concerned entirely with consumer cooperatives, is a sound guide on the general problems of Board, Member and Management relationships. The interplay of Board and Management in an agricultural cooperative has been the subject of a recent research study carried out by Wye College (*Farmers Cooperatives in Operation* by M. Brown, published by CCAHC). See also *Management in Cooperative Societies* by T. E. Stephenson (Heinemann, 1963).

7. Supply and service cooperatives

In a highly developed agricultural system such as that of the United Kingdom farmers have to rely not only on their own production skills and well-organized markets for their products, but also on the availability of farm inputs of good quality at reasonable prices. These inputs may be either the products of other farms, such as cereals, seeds and young livestock, which can be more appropriately considered in the next chapter on marketing, or manufactured goods bought from merchants or through cooperatives. In general, and not only in the United Kingdom, there is a somewhat looser relationship between farmers and their supply cooperatives than that between them and their marketing cooperatives, on which they tend to feel themselves more dependent. Because of this, the importance of the requisite cooperatives has often been somewhat underrated. Their importance ultimately turns on the answer to the question whether, in the absence of such cooperatives, competition between other suppliers of agricultural inputs would be sufficient to ensure that prices were kept down and quality maintained. The question can only remain hypothetical, but the answer given by farmers would be positive, that the competition of their own organization is essential in order to keep general trading prices and quality in line. It must be admitted, though, that a good many of them, if pressed, would as happily accept the opposite proposition, that competition from other suppliers is useful in controlling the trading policies of their own cooperatives! A sense of strong personal commitment to cooperatives supplying requisites is comparatively rare, although the farmer who spends much time in 'shopping around' is also fairly exceptional. In any case, this lack of firm attachment by members, while perhaps slowing down the rate of progress of requisite cooperatives, has at least ensured that they never cease to strive for a high level of commercial efficiency.

The total supply of manufactured inputs to agriculture (i.e. excluding seeds and livestock) in 1972–3 amounted to £1693 million, of which the main constituents were feedingstuffs 43 per cent, fertilizers 9 per cent, machinery (depreciation and repairs) 17 per cent, fuels 5 per cent, other goods and services (excluding labour) 26 per cent. As mentioned in Chapter 1, manufacture of these products has become concentrated in a fairly small number of firms.

The function of the distributor of these products is threefold. First there is the question of quality. Here the position is different today from what it was in the nineteenth century, when adulteration of farm supplies was common, particularly where fertilizers were concerned, and the early cooperative used to maintain an analyst on its staff to ensure that products bought for its members were of the quality they purported to be. Now the problem is rather that of distinguishing between qualities, with the farmer relying on his merchant or cooperative to advise him which of various rival products will give the best results in relation to its cost or ease of application. Secondly there is the question of price. Some of the earliest requisite cooperatives were known as 'truckload' societies, buying at wholesale prices consignments in bulk which were distributed there and then to members, one of whom would act as secretary and another as treasurer. A modern counterpart of these arose in the 1960s in the form of buying groups. Price improvement, however, that is to say the obtaining of retail goods at as near as possible wholesale prices, while a main cause for establishing requisite cooperatives, is seldom the sole reason for their continuing existence; most of those which have survived for any length of time have found it necessary to develop also a capacity for providing a service. This third function of a distributor has always been difficult to define, and its importance has at various times been challenged; it may be useful, therefore, to see what part it plays in the operations of a modern requisite cooperative.

PROVISION OF SERVICES

If all the members of a cooperative were equal there would be no problem in knowing what services to provide. In practice they vary enormously. There are those who do the greater part of their business with the cooperative, while there are others who use it only irregularly. There are the prompt and the slow payers. There are those who are able to take deliveries in bulk, so that a lorry does not have to be unloaded between leaving the factory and arriving at the farm store, and others who want or can manage only small lots. There are farmers who would prefer to do all their business on the telephone, and settle their debts by post, and there are farmers who welcome a visit from the firm's representative and who would never dream of writing a cheque unless he was at hand to ask them to do it. Some again are highly qualified and experienced, while others are more dependent than they would care to admit on the advice of a visiting salesman. A cooperative which follows the practice of admitting virtually all applicants to membership, provided they are genuine farmers resident in the district it serves, will be able up to a point to vary the services it provides to the needs of the particular individual. Thus it can accommodate both the man who delays payment and the man who takes only small quantities, and pass on to each of them some part of the extra charge that it incurs by providing facilities for credit in the one case and for breaking bulk in the other. It is more difficult for the cooperative to reward the customer who places his orders on the telephone and pays his cheques through the post, thus saving the very considerable expense of a visit from the representative, since the representative has to be employed in any case, where the cooperative is one which knows that the generality of its members could not in practice be relied upon either to place their orders or to settle their accounts if he ceased calling on them. Other cooperatives, however, have a special category of 'contract members' and quote special terms for them. Every cooperative has to try to establish a level of service which corresponds reasonably well with the average desires of its membership and, so far as may be possible without undue refinements of accountancy, reward the members who enable it to economize on costs and penalize the members who involve it in additional costs. Naturally, the cooperative can itself do a good deal to change the attitudes of members, when it sees these resulting in excessive on-costs, as for example by publicizing and explaining changes which the Board has decided upon in centralizing depots, altering the system of accounts, or introducing other measures designed to increase its efficiency. Whatever internal rules the cooperative lays down for itself will always be administered with some flexibility. Thus an indulgence is usually shown in allowing credit to young farmers seeking to establish themselves, or to farmers who have suffered some particular misfortune, and small farmers in general often get rather better terms than, on a strict analysis of costs, would really be due to them.

The term 'service' is also used of a cooperative in a different sense from the foregoing, to describe the level of cooperative activity, and in particular whether this activity consists of factoring farm requisites manufactured by others or manufacturing these on its own account. The principal requisite manufactured is animal feedingstuffs, though one cooperative in Scotland also manufactures fertilizers. Probably there is no question of policy more often or deeply debated in cooperative Board rooms than whether or not to go in for manufacturing feedingstuffs. In the first place there is a natural logic in doing so, for any cooperative trading in a district where a surplus of grain is produced and may in times past have been difficult to sell, but which could be sold, very likely to the same farmers who had produced it, once it had been compounded with other additives into animal feed. Secondly there is the question of cost, which debars any cooperative working on a very low margin of profit from considering this or indeed any other capital-intensive enterprise. Thirdly, there is the question of the comparative profitability of the cooperative's operations, if it carries on or does not carry on supplementary activities. Strange though it may seem, little research seems to have been done, or at any rate published, on this important question, with the significant exception of that produced by the Department of Economics of the University College of Wales, on *Agricultural Cooperative Trading in Wales*, in the late 1960s. This investigation showed that there was a strong positive correlation, in the thirty-eight Welsh requisite societies reviewed, between gross margin and final profit, i.e. the larger the gross margin, the larger was likely to be the profit. On the other hand, there was hardly any correlation between the size of the turnover and the gross margin. The universal validity of the latter conclusion is doubtful; it has to be remembered that even the largest Welsh turnover is fairly low by general standards in the UK, where it would probably be found that such a correlation did exist. The significance of the former conclusion is that any cooperative which confines itself to commodities with narrow price margins, in particular feeds and fertilizers, runs the risk of permanently low and indeed almost zero profitability, unless its turnover in these commodities is sufficiently large to enable it to extract particularly generous discounts from its suppliers. On the other hand, goods such as hardware, pharmaceuticals, implements, etc., afford a higher gross margin, and so too does the manufacture of feedingstuffs. Cooperatives may therefore be strongly induced to diversify, if they have the expertise, and to integrate distribution with manufacture, if they can afford the capital cost, so as to help carry the less rewarding activities, which as agricultural organizations serving their members they can hardly afford to discontinue. There is even one example of a cooperative having deliberately set aside a capital sum to develop a line of business in which its members have no interest whatever, in order to be able to carry on with the traditional business of feedingstuff and fertilizer supply that is expected of it.

It is difficult for a cooperative which does not manufacture feedingstuffs to escape the charge that it is just another business, hardly distinguishable from the private businesses with which it is in competition, except for being farmer-owned. On the other hand, a cooperative with its own brand of feedingstuffs is able to offer a product which is unique to itself and cannot, like the national brands, be obtained from any other dealer. Further, the reminder on the bag of the cooperative's name in the farm store, poultry house or milking parlour of every member using the product can hardly fail to result in his identifying himself with the cooperative to some extent, and in its acquiring a real meaning for him.

NUMBER AND IMPORTANCE OF SUPPLY (REQUIREMENT) COOPERATIVES

Agricultural cooperatives are classified as 'requirement' or 'marketing', though their functions overlap to a certain extent; in particular, as will be explained in the next chapter, requirement cooperatives have traditionally provided an outlet for cereals, seeds and potatoes grown on the farm. The trend over the past six years appears from the following figures:

(i) *Requirement cooperatives in 1968* (Source: Plunkett Statistics)

Country	(a) *No. of predominantly requirement cooperatives*	(b) *No. of members of* (a)	(c) *Requirements turnover (all cooperatives)* £000	(d) *Turnover in* (c) *as % of UK turnover*
England	61	136072	118347	70·4
Wales	38	50961	12637	7·5
Scotland	27	17535	6660	4·0
N Ireland	19	7659	4906	2·9
Requirements turnover of market cooperatives	—	—	10503	6·3
Cooperative companies	n.a.	n.a.	15000 (est)	8·9
	145	212227	168053	100·0

(ii) *Requirement cooperatives in 1972* (Source: Plunkett Statistics)

Country	(a) *No. of predominantly requirement cooperatives*	(b) *No. of members of* (a)	(c) *Requirements turnover (all cooperatives)* £000	(d) *Turnover in* (c) *as % of UK turnover*
England	55	131542	174149	70·2
Wales	27	43955	15742	6·3
Scotland	20	17374	11880	4·8
N Ireland	14	4750	6598	2·6
Requirements turnover of market cooperatives	—	—	14085	5·6
Cooperative companies	12	16704	26054	10·5
	128	214325	248508	100·0

The broad indications given by these figures are as follows. The number of cooperatives has fallen, mainly as a result of amalgamation; there have been very few failures. The number of members (of cooperative societies) has also fallen, possibly as a consequence of a decline in the population fully engaged in agriculture, as well as the cooperatives' policy during recent years of paying off the shareholding of inactive members, so far as they were able to do so. The relative development in the four countries has not altered significantly. Mention has been made previously of a breakaway movement in the early sixties to form 'groups', the principal characteristics of which were that they had (or were supposed to have) firm trading commitments from their members, that they operated on very narrow margins and that they were incorporated under the Companies Act. The successors to these groups are the cooperative type companies of today, by far the greater part of whose activity has up to date taken place in England, though mainly in the marketing sector.

The figures quoted above do not make clear the extent to which, as a result of the amalgamating process, the larger cooperatives have come to dominate the scene. The indications of this are as follows; the figures are again those for 1972, though in this case 'turnover' represents the total trade of cooperatives for which the supply of farm requirements is the main activity:

Requirement cooperatives in 1972, by turnover (Source: Plunkett Statistics)

	Societies		*Companies*		*Total %*	
	No.	*turnover* £000	*No.*	*turnover* £000	*No.*	*turnover*
over £10 million	7	125554	1	13000	6·0	53·8
£10–5 million	5	38655	—	—	3·8	15·0
£5–1 million	22	53081	3	3826	19·0	22·2
under £1 million	82	20526	12	2634	71·2	9·0
	116	237816	16	19460	100·0	100·0

There are some indications that the larger requirement cooperatives have a larger turnover per member. But there is nothing specially significant in this; it merely means that the larger organization has more to offer in the way of services. In any case, given the number of inactive members who are still shareholders, the relationship of membership to turnover is seldom a reliable guide.

It would appear from the summaries of published accounts that the general financial position of the supply or requirement cooperative societies has materially improved in recent years:

Net profit of requirement societies (Source: Plunkett Statistics)

	as % of turnover	*as % of capital employed*
1972	2·7	12·4
1971	2·4	11·4
1970	1·9	8·5
1969	1·5	5·9
1968	1·8	5·9
1967	1·6	5·4

It is probable that the same trend would be found in the case of the few cooperative companies engaged in the same activity.

Attempts to relate the turnover of cooperatives handling farm supplies to the value of national inputs are hampered by the fact that the latter are calculated on a June to June basis whereas the former are mostly derived from accounts based on the calendar year. Thus, to obtain a meaningful comparison, it is necessary to work out the relationship over a period of years. Unfortunately there are further difficulties. The first is that until very recently statistics have not been collected from cooperative companies at all, and they do not even now distinguish between one type of requirement sold and another. The second is that, from 1970 onwards, sales by cooperative societies have been classified differently, so that comparisons with years previous to this cannot be relied upon. Thirdly, where only global figures were quoted by the society completing the return, the breakdown has had to be estimated. With these important qualifications as to accuracy a comparison of the cooperative societies sales to producers (by financial year) with the national purchases of producers (by farming year) over the past three years gives percentage results as follows:

	1970	*1971*	*1972*
	1970–1	*1971–2*	*1972–3*
Feedingstuffs*	13·0	15·2	13·2
Fertilizers	18·5	14·6	23·8
Seeds	9·1	11·7	9·7
Fuel and oil	10·9	11·8	14·7
Machinery	6·6	6·4	7·0

* about half of which was manufactured by the cooperatives.

These figures understate the true position; if sales by cooperative companies were added they would have to be marginally increased. Unfortunately it is not possible at the moment to obtain a breakdown of these sales by commodities. The series would need to be continued for several more years before it could be said to provide any firm indication of trends.

FEDERAL ORGANIZATIONS

It will be clear from the descriptions given so far that cooperative trading in farm requirements has followed a different pattern from that found in most other countries of Western Europe, where the growth of cooperative economic influence has been marked by and, some would add, has depended upon the establishment by the primary societies of powerful second-tier or federal organizations. The reasons for the difference in patterns of development are complex and cannot be explored here. It is obvious that the early failure of the Agricultural Wholesale Society (see Chapter 2) and the availability of the services of an agricultural department in the Cooperative Wholesale Society at a critical stage of agricultural cooperative development played an important part, as did the absence of any system for determining the trading areas of agricultural societies (as had been developed, for example, between consumer societies and the Cooperative Union), which resulted in these societies regarding one another as rivals rather than collaborators. During the sixties several attempts were made by outsiders and by the managers themselves to bring about a more coordinated approach, all of which broke down on the hard fact that, in England, a number of the cooperatives had reached a size sufficiently large to enable them to obtain for themselves most of the benefits of scale so that, as far as they were concerned, federalization would not have improved their position. It would seem that there is more than a little truth, as well as jest, in the opening words of an account of how the first federal of supply cooperatives, United Scottish Farmers Ltd, came to be formed in 1971. (*Farming Leader,* September 1973.)

During 1968 Scotland seceded from the Union with England, or more precisely the managers of Scottish Societies formed their own organisation – Scottish Agricultural Managers Association. The formation of this new organization marked the beginnings of a unique pattern of development in agricultural cooperation. Many experiments in bulk buying resulted in the formation in 1971 of United Scottish Farmers Ltd as a purely commercial organization joining purchasing power of eight Scottish societies. During the first year the expectation and hopes of the founders were amply fulfilled. USF was growing rapidly in turnover and in Society members, giving Scotland the only organisation of this type in the UK.

At the outset a few manufacturers were hesitant to trade, but after a short period of initial resistance, when integrity and good will were being established, [there came] an increasing flood of offers from suppliers and manufacturers. . . .

Trading membership of this organization quickly grew to over twenty, and attracted a number of cooperatives from other parts of the country. In the following year the Scottish initiative was followed in Wales, where Welsh Farm Supplies Ltd with a trading membership of eighteen successfully completed in 1973 its first year's trading. In Northern Ireland central buying for agricultural cooperatives is undertaken by the Irish Agricultural Wholesale Society. In England, where the Cooperative Wholesale Society still provides a service to agricultural cooperatives (its agricultural and horticultural sales for the year ended January 1974 were £26·8 million) there has hitherto been no comparable development by the cooperatives themselves, although, as the establishment of Farmers Seeds Federal Ltd (see Chapter 8) has shown, even the largest of them are now beginning to think more seriously in terms of federal development.

FUTURE OF SUPPLY (REQUIREMENT) COOPERATIVES

In 1971 consultants prepared for the Central Council plans for an examination of the changes in requirements trading which could be expected in the course of the next few years, having regard to the effects of the enlargement of the EEC. In the event, following advice given by cooperative associations, it was decided to restrict the scope of the study to *Requisite Cooperatives handling Animal Feeds* – seeing that feedingstuffs alone accounted for about half the total farm expenditure on farm inputs – and a report with this title was subsequently produced. A number of the main trading organizations, including cooperatives themselves, assisted the consultants by providing information and comments.

A question with which the consultants were concerned, and one which has still not been satisfactorily answered, was whether the agricultural cooperatives generally were maintaining their share of the market. The background to this question is that farm size has been increasing, most of all in the over 500 acres category, and that pig and poultry units, which are big consumers of compound feeds, are becoming fewer in number, larger in size, and more closely integrated with both their suppliers and their customers. Having expressed their doubt whether some of the cooperatives might not, for these reasons, find themselves operating in a declining market, the consultants proceeded to examine the market shares of the national (public company) manufacturers of animal feed, which they found to account for an estimated 61 per cent of the total, compared with an agricultural cooperative share of an estimated 7 per cent. Thus there was still room for the cooperatives to expand and, in the opinion of those of them which had undertaken the responsibility of manufacturing, adequate incentives for them to do so. The consultants themselves endorsed the arguments in favour of manufacturing, and a fair part of their study was taken up with an analysis of the various factors which should be taken into account in order to make an operation of this kind a success.

It is interesting to find that both this report and the one concerned with agricultural cooperative trading in Wales referred to earlier emphasized the disadvantages, economic and political, of cooperatives acting as simple factors of feedingstuffs and fertilizers, and recommended in the one case a move from non-manufacturing into manufacturing and in the other case diversification into a greater number of lines of products in order to reduce the dependence on these two commodities. Admitting that the advice to cooperatives to extend their operations into manufacture may be sound wherever it can be followed, it is none the less apparent that, with two-thirds of manufacturing output in the hands of national firms, cooperatives will continue to have as big if not a bigger role in distributing feedingstuffs manufactured by others as in manufacturing for themselves. A satisfactory compromise between these two separate roles, in terms of target sales, for example, or related prices of manufactured and bought-in feedingstuffs, is not easily reached even where the same cooperative is handling both commodities. The conflict of interest becomes acute where two independent cooperatives are concerned, operating in adjacent and overlapping territories, one of which is committed to pushing a national brand of feedingstuffs, and the other to selling feedingstuffs of its own manufacture.

With all these competitive tensions in their situation it is understandable that the English requirement societies have found it difficult so far to evolve any common strategy in respect of the sale of farm requisites, particularly feedingstuffs, although their small minority share of this market clearly makes it imperative for them to do so. How this coordination of their policies will eventually come about is difficult to say, but it appears more likely that the formula of Farmers Seeds Federal Ltd than that of the Scottish and Welsh general purpose federals will be followed. A big and not too difficult step forward would be made if the cooperatives with the largest manufacturing interests were able to agree upon a common brand and quality, as was done with farm seeds; once such a step had been made further measures of promotion and rationalization would follow almost automatically. Other possibilities for joint action between cooperatives could be found in respect of agricultural chemicals, fuels and oils, etc. There is little doubt that some evidence of coordination along these lines, even on a modest scale, would help to improve the 'image' of cooperation in the eyes of the general farming public.

SUPPLY COOPERATIVES ENGAGED IN MARKETING

It will be seen from the table in Chapter 5 that supply cooperatives have a fairly important marketing activity, in addition to their main role. There is a question whether this activity is likely to expand or diminish in the future. It has already been noted that there are difficulties of reconciling supply and marketing objectives in the same organization, and that there is a problem too in setting up a subsidiary cooperative with separate functions, over which the parent body is unable to exercise a general control. In the circumstances one finds, as one might expect, that the marketing operations of supply cooperatives were originally conceived, and have largely continued, mainly as a service to members, are organized in a separate department which is expected to pay its way, but not to make much contribution to the general revenue. An important object has been to create goodwill for the cooperative, i.e. to help promote its sales of feedingstuffs and other farm supplies, though this may not be explicitly stated.

In principle one might suppose that such a service would be less satisfactory than one given by a cooperative which was more directly market-orientated. This is not necessarily so. There are a good many cases where the farmer has a product to dispose of which was not produced for the market, but as a by-product from doing something else. Calves from a dairy herd are an instance of this. There are other cases where, for one reason or another, farmers seem to prefer or may be forced to look to their feed supplier to provide them with a marketing service. A recently published report on *Cooperation in Pig Production and Marketing* has made the point that, over a period of only a few years, national feedingstuff firms have built up their so-called feed schemes from nothing so that they now cover about 10 per cent of all fat pigs produced. Such schemes often include a full marketing service; indeed this is an essential part, where the scheme in question involves the supply of feedingstuffs on credit, in which case it is necessary to control the final sale of the pig in order to recover the debt. Hitherto farmers' supply cooperatives, even those which manufacture their own feed, have shown no inclination to imitate this example of vertical integration, or semi-integration, although they must obviously be concerned about the possibility of improving their share of the market for feedingstuffs, or even maintaining their existing share, if they do not take counter measures of some kind.

SERVICES

Most services (e.g. transport or storage) provided by cooperatives are not separately charged for, but form part of the cost of the goods sold to the member or deducted from the price of produce sold on his account. There are other services specially provided for producers, for a charge, at their request. Many of these special services, for example that of drying a farmer's grain or spreading lime on his fields, are provided by multi-purpose cooperatives whose main function is the supply of requirements or marketing. However, there are a few cooperatives which furnish services and nothing else, such as those which provide an artificial insemination facility, a labour supply, a grass-drying operation, or nuclear stock (i.e. genetic material) for livestock or plants. It is arguable that the first and fourth of these examples ought rather to be regarded as production cooperatives, since they are essentially a grouping together of producers in order that a part of the production cycle may be performed by them more efficiently jointly than it could be individually.

In a separate category are the Pest Clearance societies, the first of which was set up in 1958 when the government decided to take advantage of the natural clearance caused by myxomatosis of rabbits, and encourage farmers by means of grants to set up local cooperatives to take permanent charge of pest control. Later withdrawal of the grants in 1971 led to a number of these societies being wound up, but in 1972 there were still 585 of them, with 43 980 members, many of them still active.

Best known and largest of all the service cooperatives is an insurance company, the NFU Mutual. As its name suggests, it is closely linked with the farmers' political organization, whose local group secretaries are the company's agents. Financially, however, it is completely independent. Its importance as an institution can be seen from the footnote to the summary of statistics at the end of Chapter 2.

NOTE: The report on *Agricultural Cooperative Trading in Wales*, referred to in this chapter, includes figures up to 1967, though most of its data are for 1965. The report on *Requisite Cooperatives handling Animal Feeds* was produced towards the end of 1972. The third report mentioned, on *Cooperation in Pig Production and Marketing,* was made to the Central Council by consultants in December 1973.

8. Marketing cooperatives

The one principle of agricultural policy about which there has, during the last half-century, been little dispute, is that farmers need to take action collectively to improve their marketing. The reasons for this view have varied from time to time and according to the different outlook of the persons expressing it. Summing up the argument, it must clearly be the aim to define marketing objectives for cooperatives which, while fulfilling the aspirations of their producer members, will at the same time contribute to the efficiency of the food industry of which the cooperatives form part, and to the satisfaction of the consumers whom they ultimately serve. A principal objective therefore, which meets all these requirements, is that a marketing cooperative should be able to bring its members' produce to the point of sale in better condition, more economically, in more convenient quantities and with greater regularity, than individual producers can manage to do for themselves. It may be, however, that other agencies can do the same equally economically, equally conveniently, etc. In that case the second objective for a marketing cooperative comes into play, which is to protect the interests of producers in relation to their market partners. Nor is this objective a selfish one. For it is often stated by food manufacturers and consumers, and their statement is to be believed, that it is in their interest that primary producers should be properly remunerated.

Government policy clearly recognizes that a proper balance of interests is desirable. In a statement made in July 1973 by the

Minister of Agriculture, after several years of deliberation on the subject, and consideration of numerous reports and memoranda, he had this to say concerning the objectives and intentions of the government's future marketing policy:

Better marketing of farm products grown in this country can help reduce costs to the consumer, provide the food manufacturers and the retailers with improved delivery of the qualities needed, and enable the farmer to grow and sell to advantage what the market requires. Since the publications of the Barker Report and the Government's Green Paper last autumn the Government have given much thought to this subject.

We believe that the time is now ripe for the further extension of cooperative and other types of group marketing among producers, and of joint ventures and contractual arrangements between producers and merchants, manufacturers and distributors and improvements in marketing more generally. If these developments are to be pursued with the necessary drive, producers must be actively associated with them....

This was followed by a further statement concerning the setting up of a new producer organization to encourage better marketing by identifying marketing opportunities and by improving methods, including, of course, cooperation among producers.

The implications of what was said by the Minister in this connection, together with his remarks about contracts and joint ventures, will be considered later. In the meanwhile it may be noted that the word 'marketing' in the statement is dependent on the definition in the Agriculture Act 1947 of agriculture as a production activity. The same definition was used in the 1967 Act, which is the authority for financial aid given to cooperatives. It covers activities which are necessary in preparing produce for market, but not any processing operation which changes the product's nature.

Leaving out of account the obligatory forms of marketing dependent on statutory Boards (possibly intended to be included in the Minister's statement, though this seems unlikely) straightforward cooperation among producers has, broadly, taken two forms, though there is increasing convergence between them. The marketing arrangements of the older established (pre-World War II) producer cooperatives catered, in the main, for the production of producers operating a number of enterprises, who were basically non-specialist. The production was not as a rule produced for any particular market; it was regarded as the duty of the cooperative to find one, and to make a better job of it than the rival merchant or auctioneer. The quality of the produce was irregular, and contracts were virtually unknown. The cooperative was expected, in the same way as the merchant, to take the market risk by buying the produce for resale. By contrast, the marketing cooperatives established since the war have tended to consist of fewer, more specialist members, with higher individual throughputs. A contractual relationship between the member and the cooperative is more common; the contract frequently specifies, not merely the quantity and time of delivery, but also production methods. Each member has a larger investment, and the cooperative often acts as a marketing agent for the producer, rather than buying as a principal. These last three characteristics, all of which imply a higher level of member commitment than in the past, have been fostered by government policy, which has made grant aid dependent on them. Such disciplines even now are not acceptable to the generality of farmers and, where the older type of cooperatives are concerned, they have to make a distinction between those of their members who will and those who will not adopt them.

Enough has been said for the moment about the general background of cooperative marketing. Now it is necessary to turn to the different sectors of production, for it is only by reviewing each in turn that an understanding of the situation as a whole can be arrived at. The object of these brief reviews will be two-fold, first to describe the present state of cooperation in the sector and then to consider the possibilities for its development. As to the latter, these will be guided but not governed by the main trends within agriculture itself. These trends were identified in a recent forecast by the National Economic Development Office for the period 1972–7 as follows:

- Continual expansion of beef and sheep production.
- A changing pattern in the production and utilization of milk for liquid consumption and for processing.
- Some expansion in the production of pigmeat and possibly of poultry meat.
- Some reduction in the egg-laying flock.

- A rationalization of and specialization in the horticultural industry.
- A limited expansion of the production of cereals and sugar beet.
- Some reduction in the potato acreage.

The actual course followed by beef production, in particular, since 1973, illustrates the uncertainty of agricultural forecasts, even when made on the highest authority.

The reviews will begin with horticulture, the most complex of the various sectors, and perhaps the most vulnerable to change. Historically it was the first of the sectors in which the policies of developing a new type of producer cooperative marketing activity were instituted, and it is consequently the one where they have had the longest time to show an effect. Where cooperative percentages of national throughput are quoted, in this and subsequent sections, the comparison will be between the 'June return' figures of agricultural output published by the Ministry of Agriculture (June 1972 to May 1973) and the Plunkett figures for produce marketed by agricultural cooperatives and cooperative-type companies for the period ended March 1973, supplemented in some instances by special surveys carried out by the Central Council. Clearly such a comparison is very far from exact, and the results must be treated with caution.

HORTICULTURE

According to official calculations for the year 1972–3, horticulture accounted for about 11 per cent by value of the total output of United Kingdom domestic produce. Rather more than half of this value was represented by vegetables (including peas and beans), about a quarter by fruit, and rather less than a quarter by flowers, bulbs and nursery stock, seeds and other minor products. In 1972–3 a series of reports on these products was produced by the National Economic Development Council, for the guidance of producers as to the lines along which production and marketing would be likely to develop in conditions resulting from UK membership of an enlarged EEC. Conclusions in these reports of relevance to cooperation were as follows:

APPLES: the best future lay with growers in favourable situations who were linked to an effective marketing system and had access to long-term controlled storage.

PEARS: as for apples, but with even more exacting storage requirements.

SOFT FRUIT: supplies for the fresh market called for well-organized distribution and marketing. (A large proportion of the soft fruit produced goes to processing.)

HARDY NURSERY STOCK AND OTHER OUTDOOR ORNAMENTALS: better organization was essential to obtain an increased share of the home market and to expand exports.

GLASSHOUSE CROPS AND MUSHROOMS: better market organization was necessary.

OUTDOOR VEGETABLES: recommendations differed with the crop, but for brussels sprouts, cabbage, carrots, cauliflowers, onions and some other crops improved presentation, storage and marketing would increase opportunities for import substitution.

The fact that so many separate reports had to be prepared shows how specialized horticultural production has become. Much of the soft fruit and vegetable production is now undertaken by relative newcomers to the industry, growing what were formerly horticultural or 'garden' crops on an agricultural or 'field' scale.

To a certain extent this development is reflected in the type of growers' cooperatives to be found. In the first place there are about a dozen older cooperatives, established mostly after but in a few cases before the First World War, which advertise themselves as handling a 'full range of horticultural crops'. Secondly there are many cooperative organizations formed since the last war, often in areas not previously regarded as having a horticultural potential, specializing in one or two crops only. The latter are usually formed of a relatively small number of members, all of whom are under contract to the cooperative, and are subject in some cases to disciplines which affect not merely the marketing but also the production operations. Such cooperatives, if not actually contracted to processors or distributors, deal with a restricted number of outlets, unlike the older cooperatives, which still tend, though to a lesser extent than formerly, to trade under a variety of arrangements in a number of the leading markets, and which in some cases run their own auctions, though auction sales are giving way almost everywhere to sales by private treaty. In a class by itself is the Land Settlement Association, a cooperative set up to administer a number of estates owned by the Minister of Agriculture, the occupants of which are required as a condition of their tenancy to abide by various rules of good husbandry, and to market through the Association. Although the individual holdings on these estates are small, their produce is much in demand on account of the consistently high quality of produce sold under the LSA label.

With one important exception, all of the main top fruit (*apples and pears*) handling cooperatives, ten in number, are grouped under a federal, the primary cooperatives being responsible for procurement, storage, grading and packing, and the federal for marketing the fruit and for ordering packing station requisites. This is a textbook example of a division of powers in accordance with operational requirements. The federal (Home Grown Fruits Ltd), while it does not, of course, physically handle the fruit, has effective control over the standards, the time when the fruit is brought out of store and the destinations to which it is sent. It has thus been able to rationalize and lower costs of distribution, while at the same time exploiting to the full the high qualities of the British eating apple in competition with other varieties. In another respect, however, the constitution of this federal is unusual, in that besides primary cooperatives it also has twenty-six individual producers as members, these being large growers with their own packhouses, whose production is as big as or even bigger than that of a cooperative. It is estimated that the entire cooperative share (by value) of total top fruit marketed (1972) was about 20 per cent; according to figures collected for the *Barker Report* they are responsible for a very significant part of the apples and pears sold under contract.

The size and success of the federal organization for the marketing of top fruit have often prompted the question whether similar second-tier arrangements could and should be developed for other horticultural crops. The broad conclusion of a Wye College study made in 1969 (*Horticultural Marketing Cooperatives: The Scope for Large Scale Organization*) was that in most other cases day-to-day selling decisions, however much they might be guided by pooled information, would have to remain decentralized with the individual managers of local cooperatives. Nothing has occurred since the report was written to cause this view to be altered.

Most *soft fruit* (strawberries, raspberries, currants, etc.) is sold to processors, and in such circumstances there is limited scope for cooperative action by growers. Eighteen cooperatives handle sizable quantities of soft fruit, but this still amounts to no more than 5–6 per cent of the total marketed.

Hardy nursery stock and other outdoor ornamentals represent a sector of the horticultural industry with considerable scope for development. So too with bulbs and cut flowers. The cooperatives' share of the 'inedible' products sector is estimated to be 2–3 per cent. Some ten cooperatives are at present involved in this business, one of which, however, claims to be the largest bulb-marketing organization in Europe.

Vegetables under glass or under cover. The main cooperative is the Land Settlement Association. The cooperatives' share of the market is significant but cannot be estimated from figures available.

Vegetables in the open. This has been an important growth area for cooperation. There are more than thirty cooperatives involved, the main crops handled being peas, brassicas, carrots, other roots and in some cases potatoes. Contracts between the members and the cooperative are usually from three to five years, the production programme being worked out between them on the basis of contracts negotiated by the cooperative with processors or distributors, and on its estimate of probable future market demand. The average capital commitment of the member of a vegetable cooperative has to be fairly heavy in order to enable the cooperative to finance a packhouse, transport and expensive capital equipment. The cooperatives' share of the market for vegetables is estimated to be 7 per cent. A recent survey of opportunities for development of cooperation in the marketing of vegetables has suggested that the operation of the packhouse is likely to be more efficient if the operator is not confined to dealing with a particular group of producers, though there will be exceptions to this rule and indeed a number already exist. If this forecast is correct the future vegetable growers' cooperatives will tend to be organizations whose relationship with packhouses will be that of contractors for the supply of produce, or joint owners of a packhouse, or owners of a packhouse which provides a service for growers outside the cooperative membership, the essence of the matter being that the packhouse must be kept working at full capacity, and must accept produce only of a packable standard.

More *peas and beans* are produced under contract than any other crop; figures prepared for the *Barker Report* show the position to be: green peas for processing 82 per cent, harvested dry 36 per cent, broad beans 71 per cent of total output. Moreover in the case of green peas for processing, which is much the most important by value of these three commodities, most of the contracts are 'transferred function' contracts, i.e. contracts where the buyer supplies some or all of the inputs, and is involved in varying degrees in production decisions. Cooperation in this sector is centred mainly around the ownership and use of expensive machinery, the mobile pea viner. These cooperatives have, as a rule, only a few members. Each cooperative has to work closely with the processing company to which it is contracted, since the margin of admissible variation from the standard laid down is very narrow. The sixty-seven cooperatives in this group are almost entirely concentrated in the eastern parts of the country, and they handle an estimated 41 per cent of the crop produced. If peas were considered separately, the percentage would be considerably higher. The interests of these cooperatives are looked after by a Processed Vegetable Growers Association, which was incorporated in 1969.

POTATOES

This crop can conveniently be considered after horticulture, since some vegetable cooperatives handle potatoes as well, and early potatoes have in the past been treated as a horticultural crop. ('Early' potatoes, lifted on or before 31st July, are a speculative crop, as well as a short-term one, and there are no cooperatives specializing in them.) Potatoes account for about 4 per cent by value of UK agricultural produce.

Since potatoes come under a Marketing Board scheme, all traders, cooperatives included, have to be licensed. Apart from this, the existence of a board makes no essential difference to their trading operation. As with horticulture, there is a distinction between the more traditional style of marketing, where the cooperative buys from its members at a firm price, and the newer style, where the cooperative sells as its members' agent. But it is no longer possible to identify the former type of arrangement with the older multi-purpose societies, since these have also in some cases adopted the newer method of trading. What one can say, however, is that the specialist, single-product potato marketing group is an entirely modern creation. A report published in 1970 (*Cooperation and the Potato Market* by E. T. Gibbons), mentions that the fifteen which were in existence when the report was written were all understood to have commenced operations during the 1960s. The author calculated that in 1967 cooperative marketing of potatoes was split equally between the specialist and the multi-purpose or multi-product cooperatives; five years later, in 1972, the latter have been overtaken.

The cooperatives' share of potatoes marketed in the same year was estimated to be 7–8 per cent. This share has been increasing fairly steadily over recent years, partly without doubt as the result of the active promotional policy which is referred to below.

A National Economic Development Office report on potatoes, published at the end of 1972, warned growers of the dangers of increasing competition from imports if they failed to improve the marketable standard. One of the measures required to bring this about was an expansion in the number of grading stations, which, if they are to be owned by growers at all, can only be owned cooperatively. This conclusion confirmed the recommendations of a study undertaken for the NFU and Central Council a year previously. This report had been quite specific about the minimum throughput (10000 tons) required to support a packing station, the operations that it would be economically justifiable for producers' cooperatives to undertake (i.e. all forms of preparation for market, but not processing), the optimum locations for the establishment of packing stations and the method of coordinating their policies. A supplementary report examined the prospects of forming specialist groups of potato growers on a semi-independent basis within cooperatives as an alternative to establishing entirely independent groups on their own, and concluded that the former policy was preferable. Here, as with vegetables, it was recognized that there were advantages in having a catchment area somewhat greater than the packhouse capacity, in view of the paramount importance of continuity of supply coupled with the maintenance of a continuously high standard.

It may be said that this report had very satisfactory results. It led to the formation of a potato-policy co-ordinating committee of the various organizations (Marketing Board, Farmers' Unions, Cooperative Associations, Central Council) concerned with fostering cooperative development. The Committee agreed on the main lines of a promotion policy and has, since it was set up, met on a number of occasions to work out the action to be taken in implementing them. This common approach has shown how much more can be accomplished in the development field when there has been clear definition of and agreement upon the objectives to be aimed at.

A question which the 1971 report had left unanswered was the scope for cooperative development within the Scottish Seed Potato industry. This was dealt with in a 1973 report which concluded that ware potato grading and marketing groups of the kind recommended earlier could undertake additional responsibilities for grading and marketing some but not all of the seed. For the rest, other organizations would be required alongside them. Such organizations, which would be on a smaller scale and with a lower throughput, might well be production companies having an additional marketing function. Finally it was suggested that, in order that Scottish seed should retain its position on the English market, it might be necessary to set up an export company, which would, however, have to be more widely based than on cooperatives alone. In fact this is how matters have turned out; in 1973 a three-member federal, Cooperative Potato Exporters Ltd, enlarged its membership to take in four private merchanting businesses.

CEREALS

Cereals – mainly wheat, barley and oats – account for about 11 per cent by value of UK farm production. No crop has been more affected as to its production by the enlargement of the Common Market. The National Economic Development Committee concluded in 1972 that the effect of this enlargement would be to encourage UK farmers to produce more cereals. This it certainly has done, with the consequence so far as cooperation is concerned that there has been some slackening of interest in forming groups for the marketing of other arable crops, which to some extent had been developed as an alternative to cereals, and a heightening of interest in grain marketing groups. Interest has also in part been stimulated by the need, which was equally noted by the NED Committee, for an increase in cereal storage capacity available off the farm. However, the economic advantage of storing grain in this way is still regarded by many farmers as marginal, compared with the traditional method of on-farm storage. A more convincing argument for cooperation where they are concerned is that the EEC regulations dealing with cereals, while underpinning the market for grain, have rendered its operations unintelligible to all but the experts. In these circumstances many farmers have made up their mind that some of these experts should be employed in securing a good price for producers, rather than in procuring grain as cheaply as possible from producers. This has provided a new reason for cooperation.

The marketing of grain in the United Kingdom has in the past always been closely associated with the supply of seeds and fertilizer required to produce the crop, and of the animal feedingstuffs into which a substantial part of the grain was converted. When Professor Britton carried out his survey of *Cereals in the United Kingdom, Production, Marketing and Utilization* for the Home Grown Cereals Authority, farmers questioned by him mentioned reciprocal trading as the most useful of all the services provided by the grain merchant. Consequently, among the agricultural cooperatives, it was the requirement societies rather than specialist marketing societies which bought the grain from farmers who were prepared to sell it to them. The business was always very competitive, and sales of grain were normally excluded from bonus. It was, however, a fairly well-honoured tradition that whoever bought the grain would also be given the order for fertilizers and feedingstuffs, or at least a part of it. Professor Britton's sample, based on the 1966–7 harvest, showed that cooperatives at that time accounted for only 6 per cent of total grain purchases. In the same survey specialist grain merchants – defined as those with three-quarters or more of their total turnover in sales of straight grain – accounted for 22 per cent of grain purchases.

Reciprocal trade in grain and requisites is still extremely important but it is clear from recent developments that straightforward grain marketing is growing in favour. The indications of this are not merely that the newly formed grain marketing cooperatives are all specialist organizations without other functions, but that a number of the requisite or multi-purpose societies have, while continuing to buy grain in the traditional manner, set up grain groups among their members to sell on their behalf – and no longer as principals – the stocks of cereals committed to them. It is an essential feature of all these organizations, whether formed under the aegis of multi-purpose cooperatives or independently, that the commitment should be by way of contract, so that the skills of the marketing executive can be fully engaged on the job of marketing the grain, and not wasted on finding out whether the grain is still there at the time when he wants to sell it, and whether its owner is prepared to accept the price offered.

The multi-purpose cooperative which sets up these grain marketing groups has to make it absolutely clear that its grain buyer, procuring raw materials for its feedingstuff operation, will only be able to buy from the grain marketing officer employed on behalf of the group if his buyer's price is fully competitive. Equally, the grain marketing officer cannot expect any favoured terms from the grain buyer. This is an interesting situation, for it means that, to satisfy the grain producer that he is being given a fair deal, the cooperative must break the chain of vertical integration which it had established within its business, and make the operations of grain purchasing and feed manufacture completely independent of one another.

As in the case of potatoes, the strategy for developing grain marketing has given rise to much concern. There are numerous areas of possible conflict. Specialist grain groups, employing their own marketing managers, have been set up in the trading areas of existing multi-purpose cooperatives, whose reaction has been to set up groups of their own, also employing a marketing officer. Other groups of farmers, who had been accustomed to sell their grain to a specialist grain merchant (as defined in the Britton report) have been encouraged by these merchants to organize themselves into marketing groups, for which the merchant acts as managing agent. From the various studies that have been made of this problem two main facts have become clear. The first is that grain marketing has become a highly specialist function, and small groups are not likely to recruit, or if they do will not retain the services of an expert, because their throughput will not be adequate for his salary. Some rationalization is therefore to be expected of the cooperative groups handling grain, in the course of which much of the present rivalry will be eliminated. The other conclusion is that the formation of a cereals federal on a national or regional scale would not be a feasible proposition in the short term, though such a federal may develop when a sufficient tonnage of grain is being handled cooperatively. The cooperative share is growing; in 1972 it was estimated to be 14 per cent of grain sold off farms.

One of the reasons for rejecting the idea of a federal at this stage is that some of the functions which it might be expected to perform are being carried out by an international cooperative-owned company, Eurograin Ltd, or rather by its subsidiary, Eurograin (GB) Ltd, which is located in the United Kingdom. This is a grain brokerage organization, however, and only a small part of its business is done with cooperatives. The UK shareholding in both companies is held by Farmers Overseas Trading Ltd, a cooperative federal formed in 1967, of which twelve of the largest multi-purpose (and grain handling) cooperatives in the UK are members.

OTHER ARABLE, FORAGE AND ORCHARD CROPS

Cooperation in *sugar beet* is confined to production, since the British Sugar Corporation is the only buyer in the market.

Hops also are, under the Hops Marketing Scheme 1932, marketable only to the Hops Marketing Board. Proposals for cooperative hops processing have not so far got beyond the planning stage.

The production of *maize* for grain is still in its early stages in the UK. Two specialist cooperatives for production and marketing have been formed. Maize production for forage does not justify separate marketing arrangements.

One cooperative has been formed for the marketing of oilseed *rape*; this cooperative has combined with other cooperatives concerned with but not specializing in the crop to form a marketing federal, United Oil Seeds Ltd.

Two cooperatives have been formed for the drying and marketing of *grass* produced by their members.

There has been a growing interest in the making of *wine* from grapes grown in favoured sites in the south and east of England. A cooperative, the English Vineyard Association Ltd, has been formed to promote the interests of the growers.

Over a hundred *Women's Institute Market* societies exist for the sale of their members' home and garden produce.

SEEDS

The pattern of trading for cereal and herbage seeds is similar, that is to say both are bought from and sold to farmers. They therefore appear twice in the national farm accounts, as 'intermediate output' on one side and as 'input' on the other. The proportion of total agricultural output which they represent is negligible, being not much more than 1 per cent by value.

Seeds, like feed (another intermediate output), have always been part of the stock-in-trade of the agricultural merchant, or requisite cooperative, for whom the crop is often grown under contract. It is a crop with good possibilities for cooperation, in view of the testing, cleaning and dressing which is required to put it in a marketable condition. The enhanced opportunities for cooperative trading which the Common Market presents have been taken to form Farmers Seeds Federal Ltd, with a trading area over the whole of England, Wales and Scotland. Twenty-six cooperatives are in membership of this federal, which accounts for 1¼ million acres of cereal seeds. Initially the federal handles only cereal seeds, but it is likely to expand into herbage seeds, which at present are handled by cooperatives individually. It is estimated that 15 per cent of the seed produced for sale is cooperatively handled.

CATTLE AND CALVES

Fat (i.e. finished) cattle and calves accounted for 18 per cent of UK agricultural output in 1972. This figure naturally disregards any transfer of stock from farmer to farmer during its lifetime, although for the farmer making such a sale the animal sold forms part of his agricultural output and the selling operation constitutes a marketing activity.

In mid 1973 the National Economic Development Office produced a report on the problems arising for beef as a result of the enlargement of the Community. It is significant that all the problems mentioned related to production, and that marketing was not discussed. In fact there is not much that the marketing of a live animal can do which adds significantly to its value, and consequently the scope for cooperation in this sector is somewhat limited.

Cooperation in respect of the semi-finished animal is centred round the marketing of store cattle and hand-reared (twelve weeks) or suckled calves. The function of the cooperative here is to pass back to the member information about market requirements, having first translated these into production terms which the farmer can understand. Secondly, it can help to ensure that stock are made available to buyers in homogeneous lots. Thirdly, it can develop regular market outlets, thus saving the time of both producer and buyer, and reducing transport costs. These may appear to be fairly modest aims, but they can, if energetically pursued over a sufficient period of time, make a useful contribution to marketing returns. The cooperative concerned may be a specialist beef or calf organization, or a general livestock organization, or a livestock department within a multi-purpose cooperative.

The above forms of cooperation overlap to a considerable extent with the cooperative marketing of finished stock. Thus a cooperative handling calves may sell them for slaughter or for further rearing, according to the state of the market. The same cooperative may sell its members' stock as finished or semi-finished, depending on circumstances. All one can say is that there is a broad distinction between cooperatives selling finished calves or cattle by private treaty, and those selling stock by auction, though in the latter case it is difficult to distinguish statistically between stock which has or has not been finished. Cooperatively owned auction marts are particularly important in Scotland and Northern Ireland.

All these uncertainties make it very hazardous to estimate what proportion of the Ministry figures for fat cattle and calves can be attributed to cooperatives, omitting from the latter any intermediate sales. The roughly estimated cooperative share of 7–8 per cent does not really do them full justice, since so much of cooperative livestock marketing activity is concerned with the semi-finished animal, which is excluded by definition. If one were instead to assess the proportion of cattle and calves which at some point in their lifetime have passed through cooperative hands, it would be much higher.

Finally, it will be noted that the Ministry figures quoted in the *Annual Price Review* all refer to livestock and not to carcasses or meat. This is because slaughtering is officially regarded as an industrial and not an agricultural activity in the United Kingdom. Farmers themselves do not accept this, but regard slaughtering as a logical sequel to their production operations, and a stage at which much of the added value occurs. There are a number of farmers' cooperatives owning or renting an abattoir, or slaughtering under contract. Those which handle cattle and calves mostly also handle other stock as well. The main producers' effort in this field has gone into the setting up of the Fatstock Marketing Company, now trading under the name of FMC Ltd and a public company, but one which is still very largely farmer-owned.

SHEEP AND LAMBS

Sheep and lambs accounted for 4 per cent of UK agricultural output in 1972. As in the case of cattle, the figures quoted in the *Annual Review* White Paper are those of final sales. The value of lambs sold off the hills for finishing in the lowlands is therefore disregarded in these figures, though for the hill farmer they constitute his main source of income.

The functions of the sheep or, more usually, lamb marketing cooperative are the same as those described under the last section, namely to convey information back to producers that will enable them to meet market requirements, to arrange level batches, to develop market outlets and to rationalize transport. It may be said, however, that the cost savings and sale premiums amount to a larger part of the final price in the case of sheep than they do for cattle, and the stimulus to form cooperative marketing groups has been correspondingly stronger. The cooperatives may be specialist, or general livestock organizations, or departments within a multipurpose cooperative.

For descriptive purpose it will be useful to distinguish between cooperatives performing an intermediate and a final marketing role. In the former category come certain cooperatives, examples of which are to be found in England, Scotland and Wales, which specialize in producing half-bred ewes from pure stock, or in some cases pure-bred ewes, which are then sold to farmers for crossing with specially selected rams, in order to produce lambs with good carcass characteristics. The sales may be 'sponsored sales' taking place at advertised centres, or sales by private treaty. Again in all three countries there has been an upsurge of interest in store lamb marketing groups, which are able to guarantee buyers stock of uniform standards, to reduce transport and marketing costs, and to give producers more security and a better return.

However, it is at the final stage of marketing that the cooperatives have made their biggest impact. The aim of the fat lamb marketing schemes is to make available large numbers of stock of a known type, weight and quality. The most successful ventures so far in this field have taken place in Wales, where eight county quality lamb cooperatives market over 100000 head of stock annually through their federal, Welsh Quality Lambs Ltd. It is in fact the success of the fat lamb marketing schemes which has led to similar schemes for store lambs.

The lamb marketing cooperatives would normally handle culled ewes as well. Lambs are also marketed by non-specialist cooperatives, including, of course, the cooperatively owned auction marts, and are sold direct to cooperatively owned slaughterhouses. It is estimated – but only a rough estimate is possible – that 10 per cent of the national production of fat sheep and lambs is marketed cooperatively.

PIGS

Fat pigs accounted for 12 per cent of the value of farm output in 1972. A report on pigs and pigmeat in that year by the Economic Development Committee for Agriculture, prepared with the object of giving guidance on the lines on which production and marketing might develop in an enlarged Community had, surprisingly, nothing to say concerning marketing, although there is no commodity more deeply affected, as a result of the necessity to dismantle the elaborate system of control over prices which had existed prior to that time.

The special feature distinguishing the production of pigs from that of other large animals is the proportion of the cost of food (70 to 80 per cent) to total costs. The corollary to this is that a substantial proportion of feedingstuffs manufactured (about 25 per cent) consists of pig rations. Pig producers and feed manufacturers are therefore and will remain to a high degree interdependent.

With pigs as with other forms of livestock there is a distinction to be made between intermediate and final marketing. Cooperation has hitherto been more developed in the former sector. Within this are found a number of cooperative breeding groups which have been established in order to supply commercial pig breeders with improved stock. However the commonest type of cooperative is the Weaner Group. Before these groups were organized – either as separate bodies or under the aegis of multi-purpose cooperatives – the producers of weaner pigs used to market them haphazardly through local markets, a system which led to unpredictable price fluctuations as well as a far from negligible disease risk. Under the weaner group system prices have become much more consistent, transport and overall marketing costs have been reduced and there has been a steady improvement in standards. Cases have occurred where the weaner producers concerned, wishing to expand but doubting the capacity of existing outlets to cope with their extra production, have set up a central cooperative fattening unit for this purpose, which is in turn contracted to a bacon factory or slaughterhouse. (The reverse situation is also found, where pig fatteners with an inadequate supply of weaners have cooperated to set up their own breeding unit.) At that stage the cooperative becomes involved in final, as opposed to intermediate, marketing.

With weaner groups set up under multi-purpose cooperatives in particular it is more and more common for the weaner-feeder arrangements to be complemented by a marketing service to feeders for the finished pig. Some specialist cooperatives also take their members pigs right through to market, while other pigs are sold through cooperative auctions, and to cooperatively owned slaughterhouses or bacon factories. Any estimate of the proportion of pigs marketed cooperatively can only be roughly calculated, but it may amount perhaps to 5–6 per cent of the total. For reasons already explained, this figure excludes pigs marketed intermediately. There can be no cooperative marketing of pigs in Northern Ireland, where a Pigs Marketing Board is the sole buyer.

A study of the future policies open to cooperatives in the sphere of pig production and marketing was commissioned by the Central Council, having been sponsored jointly by the Meat and Livestock Commission, Farmers' Unions, Cooperative Associations and other organizations. The report appeared in 1973. Its principal theme was that cooperatives should seek to develop, rather than avoid, the interdependence of producers and feed suppliers already mentioned, and an equally great interdependence between producers and processors which was likely to be the long-term consequence of the abandonment of the former price-support system. Another reason suggested for this closer integration was that it would be more logical, under these new conditions, for pig meat rather than the live pig to become the marketable commodity, in respect of which transfer of ownership would take place.

POULTRY

Poultry account for about 6 per cent of UK agricultural output. Production is now virtually entirely in the hands of specialist producers and on a factory scale. Cooperatives play only a small part in poultry marketing.

MILK AND MILK PRODUCTS

Milk and milk products are by far the most important item of UK agricultural output in terms of value, accounting for 22 per cent of the total. As a result of the Marketing Acts of the 1930s, voluntary cooperatives have very little influence in this sector.

In Scotland the cooperative dairies were absorbed into the Boards when the three Milk Marketing Schemes were introduced in 1933 and 1934. In England and Wales most of the cooperative dairies which existed in 1933 have been closed down or taken over, with the exception of one cooperative creamery in Wales and one cooperative cheese-making plant in England, the former acting as an agent of the Milk Marketing Board and the latter receiving a milk allocation. Only in Northern Ireland are cooperative creameries still important. The total cooperative share of milk and milk products marketing is consequently very small, amounting to not more than 1–2 per cent.

EGGS

Eggs account for about 7 per cent by value of UK agricultural production. Immense charges have taken place in the marketing of eggs in recent years, hastened by the winding up of the British Egg Marketing Board in 1971. Formerly a substantial part of eggs handled cooperatively used to pass through the egg departments of multi-purpose cooperatives. Now, however, that the general farmer has practically gone out of egg production, leaving it to specialists, cooperative egg marketing is practically confined to cooperatives having this single function. Four of these have combined to form a federal marketing cooperative, Goldenlay Ltd, which has close trading links with the supermarkets through which the bulk of eggs are now sold.

The cooperative share of eggs sold off farms (only a proportion of which pass through packing stations) is calculated to be 15–16 per cent. A study of the role of cooperation in the egg industry produced for the Central Council, the NFU of England and Wales and the former Agricultural Cooperative Association concluded that the future lay with large-scale units in which feed production, egg production, packing and marketing were under central control, and would form a consciously planned single unit. This degree of integration is still very far from having been achieved as yet in any egg cooperative, but with the growing recognition of the identity of interest between egg producers and the egg packing cooperatives of which they are members progress is all the time being made in that direction.

WOOL

Clip wool accounts for less than 1 per cent by value of UK agricultural production. All such wool has under the 1950 Scheme to be sold to the British Wool Marketing Board, for which farmers' cooperatives, specializing in wool but usually handling as well the particular requisites needed by sheep farmers, act as agents, along with other firms, in some of which the Board itself has a financial interest. It is estimated that the cooperative share in the handling of wool amounts to 32 per cent.

In conclusion, something must be said about the general strategy of developing cooperation in marketing, and particularly about the alternatives of building on existing organizations or initiating new ones. In principle, the existing organization must obviously be preferred, since not only will overhead costs be reduced if this is done, but, more important, one body of farmers will not be set in competition with another to lower the price that both receive. Notwithstanding these obvious advantages of working together, rival organizations continue to exist and are constantly being created. Sometimes this situation is due to basic disagreement, too wide to be bridged, about policies and methods of operation, but just as often to personality differences or private ambitions. In the United Kingdom where, unlike many other countries, there is no legal obstacle to cooperatives with similar functions trading in the same area, it is sometimes proposed that other forms of discouragement might be used, in particular the withholding of grant aid, where this is available on a discretionary basis, from cooperative organizations which would duplicate facilities already existing in the area concerned. It is interesting to compare this proposal with one made under the EEC common agricultural policy (in connection with producer groups) that 'grants shall not be made if the operators in a particular region already have at their disposal sufficient of the plant. . . .' Such a provision, well meaning though it is, seems likely to lead to endless argument between those already in possession and the applicants for grant, the latter protesting that the existing plant is inefficient, or badly managed, or not available to them on fair terms. However independent and careful the judgement on such issues, it can hardly fail to be unpopular with one side or the other.

NOTE: In addition to the reports mentioned in this chapter there is a wide variety of pamphlets, produced by the Central Council for educational purposes, which briefly outline different types of marketing cooperatives, and the factors which were taken into account in setting them up.

9. Cooperation in production

As was mentioned in Chapter 7, there is some overlapping of function between cooperatives providing supplies or services and cooperatives concerned with production. In view of the difficulty of precise definition, it may be useful to list various forms of co-operative production activity which occur in the United Kingdom, roughly in order of the intensity of the cooperative members' in-volvement in them. The categories listed correspond fairly closely with those given in a recent EEC publication describing forms of collaboration in agricultural production in the 'Six', though with some differences of emphasis.

The list does not include the case of a production unit which is owned by a producers' cooperative, but which is not directly related to the production activity of its members. An example might be an experimental farm established by a supply cooperative for the testing of its manufactured feed, or a production unit set up by an egg-marketing cooperative to fill a gap caused by a shortage of supplies in a particular area. Such activities as these, however essential they may be to the cooperatives concerned, cannot themselves be regarded as being of a cooperative nature.

The EEC list recognizes four forms or grades of cooperation in production:

(i) Utilization in common of one or more means of production. The sharing may be of labour or it may be of equipment. The use to which the shared resource is put on each member's farm re-mains his individual responsibility.

Examples in the United Kingdom are agricultural labour camps and machinery syndicates. The latter are discussed later in this chapter.

(ii) Simpler forms of cooperation involving profit sharing or equalization between a number of farms, by means of arrangements which permit a division of labour between farms. A distinction is made in the EEC text between crop and livestock production. In crop production such an arrangement might consist of one farmer undertaking the production of seeds or root-stocks on behalf of the rest of the group. In livestock production an example would be a group consisting of a calf rearer and a certain number of milk producers, or a group of pig fatteners and a number of weaner producers. It is an essential condition that there should be some sharing of the risks; if, therefore, financial responsibility were transferred at the time of the transfer of the stock, there could not be said to be any true cooperation in production.

In the United Kingdom examples of this kind of cooperation exist, but mainly on an informal basis. In principle, there seems to be no valid reason for distinguishing between this and the next kind of cooperation in production to be mentioned.

(iii) Partial integration, or the operation by farmers in common of one branch of production, the independence of the member farmers in the group continuing unabated in all other respects. Cooperation of this kind would normally centre around a particular farm enterprise, which is operated under an agreed set of rules and possibly under common management. It may also include arrangements for extending the production chain forwards or backwards by means of the joint activity. Again a distinction is made between crops and livestock.

This kind of cooperation is also common in the United Kingdom, and may be informal or formal. Examples in respect of crops are pooled arrangements for the production of a particular crop, such as sugar beet, sprouts, peas, etc., by general farmers, or cooperation in one phase of their production, such as harvesting. Producers may also make arrangements for extending forward the production chain, as, for example, by centralizing the grading of potatoes (which may, however, be the first link in the marketing chain if, having graded the produce jointly, the members also join in the selling of it.) The process is extended backwards if they combine in order to produce seed or rootstock, as in the nuclear stock associations formed for the production of strawberries, raspberries, ornamentals, top fruit and rhubarb, although these have sometimes been regarded as requirement or service rather than production cooperatives, possibly because they are on too large a scale for the individual producers to take an active part in them. Examples in respect of livestock would be groups for the development of breeding or fattening units in connection with members' herds, or possibly the administration of common grazing land. Artifical insemination cooperatives may be regarded as a form of cooperation in production under this heading, though these too have tended to be classified as supply or service cooperatives, again because of their comparative remoteness from the farm operation.

(iv) Complete integration, in which all the production enterprises on the members' farms are controlled by them jointly.

Very few full-farm integrations have taken place in the United Kingdom, and even fewer have been successful. It is unlikely that, in the absence of special incentives, they will ever be numerous. The reasons for this are partly psychological, partly legal (as will be explained later), and partly that mixed farming systems, which predominate in the United Kingdom, are likely to be more difficult to integrate successfully than monocultural ones. In any case, it seems that a full-farm integration, carried out so as to take full advantage of the assets that have been pooled, ought to be regarded as a first step towards amalgamation rather than as a final step in cooperation, since the more complete the integration, the more difficult it becomes to reverse it. Partial farm integration, on the other hand, can be planned from the outset as a long-term but impermanent arrangement, which can still become permanent, if found to be clearly in the interests of the producer members – and their successors. This leads to a further point of definition. Cooperation in production is an arrangement whereby several farms combine for the joint management and operation of some or all of their inputs, including land. There is no question, in such an arrangement, of collective ownership of land. On the contrary, if cooperation involves a voluntary act of association, there must equally be a freedom to withdraw from that association. Obviously the higher the degree of cooperation, the more difficult withdrawal becomes; but the possibility must always exist, or else the act of association becomes one of amalgamation.

THE CASE FOR COOPERATION IN PRODUCTION

The cooperation in agricultural production being considered here is quite distinct from the practice of good neighbourliness and mutual help in emergencies, for which farmers are as renowned as they are for their independence in normal times. It involves a formal relationship, in which some amount of independence has to be surrendered, in return for other advantages. Cooperation of this kind is a fairly recent development, which has made significant progress in Western Europe in recent years, particularly among more advanced farmers, who have been quicker to recognize its economic possibilities, and less inhibited perhaps by past tradition. A publication which summarises the lessons of their experience, though it does not draw on British examples, is *Group Farming*, produced by the Organization for Economic Cooperation and Development, the English translation of which appeared in 1972. It deals entirely with partial and complete farm integration. The practice of group farming has both economic and social aspects. On the economic side, better use can be made of inputs and work can be more efficiently organized, through each one of the members being able to specialize and concentrate on the job in hand, free from distraction and worry about what may be going on at the other end of the farm. Also it becomes possible to use the right machine for the job, instead of one which is either over- or under-powered. After an initial running-in period of two or three years, when the farm structure is still being reorganized, new capital commitments are being incurred, and the members of the group are getting used to working together, they can look for higher incomes as a result of lower costs and, so far as can be judged from results in France and Spain, these group returns are equal to and usually higher than those for comparable individual farms. However, the report says, it is difficult to be sure whether this is because the members of the group to some extent represent an élite in the agricultural world or whether the possibilities for specialization offered by group farming have been responsible for a decisive improvement in the technical skill of the partners. This improvement is likely to be progressive. Following the rationalization of production operations, available labour will no longer be fully used, and one of the ways of dealing with this situation is to intensify production.

Another result of cooperation in production is to increase the amount of leisure, to put an end to the heavy work to be done by women, and to make it possible for the men to have days off from the farm and holidays. In a full-farm integration, if one of the men in the group falls sick, his source of income is not seriously endangered. This greater security may lead to a more dynamic attitude among the partners in that their feeling of solidarity may induce them to take certain risks which they would not dare to take alone. Also, the practice of working together must give each of the partners a far better understanding of management and work organization than he could have attained singly. But, of course, the psychological difficulties in the way of this kind of cooperation are prodigious, particularly in the case of a full-farm integration, when the possibilities of an easy retreat in the case of disagreement or incompatibilities are no longer available. Group farming is, therefore, something for the younger rather than the older farmers, and, in British experience, is better approached by way of a partial farm integration, during the course of which the members of the group can find out from the experience of working together whether they want to take it any further.

Another possible result of cooperation in production is to make it possible for one of the members of the group, whose stake in farming is too big to abandon but too small to provide him with a full livelihood, to move first into part-time work and later into full-time work outside farming.

SOME PROBLEMS

Cooperation in production can be regarded from two different points of view. It may be seen as a temporary, even if fairly long-term, arrangement, which can be brought to an end if there is any fundamental change in the stakes or attitudes of the partners, or an arrangement which is expected to be permanent. A full-farm integration will require investments to be made in buildings, etc., on the land of one of the partners, and it is desirable therefore that the group, for its collective protection, should have at least a lease over the land concerned, if not over the whole land which is being farmed jointly. However, under English law, such a lease creates an agricultural tenancy, which cannot thereafter easily be revoked. The member of a farming group will be understandably reluctant to sign such a lease and, if he is a tenant farmer, he will not be permitted to do so. This is a legal difficulty to which no solution can be found under the law as it stands at present, and it forms a decisive obstacle to the development of full-farm integration in the United Kingdom. An incidental difficulty is that the group cannot offer land or buildings as security for loans, but must rely on the joint and several guarantees of its members.

The legal form adopted for most cooperative production activities in the United Kingdom is a partnership. The rights and duties of the members are determined by the partnership deed and follow a fairly standard pattern which experience has shown to be fair and practical. Capital contributions of the members are rewarded at a rate of interest which is not higher than that approved by the Registrar in the rules of an incorporated cooperative society. Services provided by a member or members are paid at the rate for the job. The residual surplus may be divided either on the basis of the yield from the acres put in by the different partners, in the case of field crops, or, preferably and more usually, in accordance with a predetermined ratio.

The fact that production cooperatives are almost all unincorporated organizations which are also unregistered, except in so far as they may have registered a business name (which is not essential), makes it impossible to have any certain knowledge of how many there are or what volume of production they account for. However, it may be of some interest to mention what are the main types of organization and in what sectors of agriculture they have been predominant.

PRODUCTION SYNDICATES

These groups, which fall within category (i) of the EEC list, have been labelled syndicates in order to distinguish them from production groups proper in categories (ii), (iii) and (iv). Membership does not involve any important joint activity outside joint ownership of the asset, meeting whatever costs are related to it, and deciding how the use of it is to be shared. Production syndicates are, in other words, a low-intensive form of cooperation, but one nevertheless carrying obligations which need to be formalized. The best-known examples of production syndication in the United Kingdom are the machinery syndicates first developed in southern England in the early 1950s – as distinct from the informal machinery sharing which, according to a small survey carried out by the Agricultural Development and Advisory Service in 1970 (see *Machinery Sharing in England and Wales* – MAFF Farm Mechanization Studies No. 19), is quite widely practised. The basis of this formal system was the establishment, in the majority of English and Welsh counties, of County Syndicate Credit Companies, limited by guarantee, through which applications from groups of farmers ('machinery syndicates') wishing to combine for the purpose of operating machinery jointly were forwarded to one of the banks which had opened up a special line of credit for this purpose. The initial screening of applications by the county companies, together with the acceptance by the syndicate members of strict rules relating to maintenance of the machinery to be purchased as well as to its day-to-day operation, made it possible for these syndicates to borrow on specially favourable terms, to repay by relatively easy instalments, and to avoid having members' individual sources of credit in any way compromised. The general experience of all those who have participated in machinery syndicates is that virtually none of the problems which might be foreseen in machinery sharing were in fact encountered; but all the same the system has so far failed to win the approval of farmers on any general scale, as was at one time hoped. The county organizations in England and Wales are loosely organized into a national Federation of Syndicate Credit Companies. Scotland has a similar scheme which operates through Agrifinance (Scotland) Limited.

CROP PRODUCTION GROUPS

The term is used to describe groups of farmers falling within category (iii) or, in some cases, category (ii), who share the use of machinery and, in addition, agree to commit a certain acreage to the group and to accept a common policy in respect of cultivation, fertilizer application, harvesting, etc. Substantial encouragement was given to the formation of such groups between the end of 1969 and April 1971 when the Central Council was able to recommend that any such groups, which accepted its standard rules, should receive capital grants in respect of assets purchased by them, including machinery already owned by a member. Subsequently this facility was withdrawn, when it was decided that production groups should be brought into line with individual producers so far as the eligibility of their equipment for grant aid was concerned. During the period for which these working capital grants were available, considerable interest was shown by farmers in forming groups for production purposes, as can be seen from the following figures extracted from Central Council reports:

Production Groups Receiving Grant Aid on Projects
(not feasibility studies or development and research)

	No. of groups	*Membership*	*Average membership*
Fruit	3	19	6
Vegetables	11	60	5–6
Peas	59	605	10
Potatoes	11	43	4
Sugar Beet	16	74	4–5
Cereal	2	24	12
Grass	271	853	3
Production/integrated	10	51	5

LIVESTOCK PRODUCTION GROUPS

These groups also fall within categories (iii) or (ii), although in most of them the importance of the common marketing element outweighs that of the common production element. There are three aspects of cooperation which are commonly stressed. The first is genetic improvement. This may be attained through membership of a large artifical insemination cooperative or by joint purchase and use of bulls, boars or rams by the group. Secondly, there are various forms of common services. These may involve a production programme jointly planned by the members; for example, if the group is concerned with producing weaners for finishing, the members must be able to guarantee to take up the weaners produced, whereas if the aim is to finish pigs, they must be able to supply the stock to fill the places available. A different form of common service is where the group employs an adviser to help its members to improve the quality of their stock. Thirdly, examples may be found of beef and sheep production groups, in which land for grazing, machinery for cultivation of most crops, and livestock are committed by the individual members; perhaps the most interesting examples of such groups are those whose membership is composed of hill farmers who have the facility for producing store stock, and lowland farmers, who have the facility for finishing them. A growing number of livestock owners, particularly in the hill areas, are involved in some sort of cooperative production activity, but statistically this cannot be separated from the marketing activity. It will have been noted under the heading of crop production groups that a great many of them are centred round grass production (usually hay or silage). The members of such groups would in nearly all cases be livestock farmers, and this may be the best form of cooperation open to them.

EFFECTS OF EEC MEMBERSHIP

With effect from 1st January 1974, the Common Agricultural Policy was brought into force in the United Kingdom in respect of aids to agricultural production through the application of Directive EEC/72/159 (Modernization of Farms). This was implemented by way of the Farm and Horticulture Development Scheme (Statutory Instrument 1973, No. 2205) under which producers can if they so wish apply for aid in respect of a development scheme jointly undertaken no less than aid in respect of individual schemes. In practice, few of them are likely to do so, because of the complexity of the procedures and the restrictions on eligibility.

However, the Modernization of Farms Directive also contained one Article which applies to farmers generally. Its aim is to encourage the establishment of recognized groups whose aim is 'mutual aid between farms, a more rational pooling of agricultural equipment, or joint farming'. These terms are fairly vague, and the precise circumstances that they cover can only be worked out by a process of trial and error. An important provision is that the aidable costs are costs of management only, which costs are available to UK production cooperatives in any case, under the national Agricultural and Horticultural Scheme 1971. The only extension of existing arrangements, as a result of the introduction of the Common Agricultural Policy in the production sector, is that non-cooperative groups are eligible for aid as well as those which are cooperative. Following the implementation of the Modernization of Farms Directive, production schemes generally are taken outside the scope of aid under Regulation EEC/17/64, the use of which is now confined to agricultural marketing projects, as explained in Chapter 4.

NOTE: Two publications dealing specifically with the problem of cooperation in agricultural production in the United Kingdom are *Cooperation in Farm Production* by G. H. Camamile, published by the Agricultural Cooperative Association in 1968, and *Joint Enterprises in Farming,* published in the same year by the Country Landowners Association. Model rules have been published by the Central Council for use by Forage Conservation Groups, which are adaptable for use by other cooperative production enterprises. The EEC publications referred to in this chapter is *Nouvelles formes de collaboration dans le domaine de la production agricole*, published in 1973 as No. 110 in the series *Internal Information on Agriculture.* The OECD publication referred to is No. 51 72 01 1 in its catalogue.

10. Central Associations and Central Council

When a sufficient number of primary agricultural cooperatives have been formed, and become successfully established, it is natural for them to want to set up a federal or second tier organization, for mutual aid and protection. In the United Kingdom, however, the initiatives to set up the Irish Agricultural Organisation Society, established in 1894, and the Agricultural Organisation Society (of England and Wales), established in 1900, seem to have come mainly from outside the cooperatives, and in the former case actually to have preceded them. Both bodies originally included in their membership a large number of private individuals, who were interested in these societies as missionary bodies for the extension of cooperation in areas and activities where it had not so far developed. In Scotland, even to this day, the Scottish Agricultural Organisation Society, formed in 1905, still has several hundred individuals as members, who are prepared to pay an annual subscription for the privilege of belonging to an association with whose aims they are in sympathy, though their influence on its policies is now very slight. In England, Wales and Northern Ireland the members of the central cooperative associations are all or nearly all corporate organizations, i.e. cooperative societies or companies. All four associations have development functions, concentrated chiefly in the marketing sector. Alongside these there have grown up the more typically 'trade association' functions of providing central services for members, of helping them to work out common policies, and of representing their interests to government or other authorities. The bodies carrying out this primarily political role are, in England, Agricultural Cooperation and Marketing Services, and in Wales, Scotland and Northern Ireland, the Welsh, Scottish and Ulster Agricultural Organisation Societies. These four bodies have delegated some of their responsibilities, in particular those concerned with external relations, to a Federation of Agricultural Cooperatives of the United Kingdom.

ROLE OF THE COOPERATIVE CENTRAL BODIES

The non-development functions of the central cooperative associations in the four countries consist mainly in their provision of

- common services, to the extent that these may be required, such as market intelligence, advice on financial and legal problems, constitutions and contracts, or assistance with public relations. Unlike central bodies in some other countries, they have no statutory auditing responsibilities. Another point of difference is that agricultural cooperatives in the United Kingdom habitually belong to general trade organizations, as for example the association of Seed and Agricultural Merchants, of Egg Packers, of Machinery and Tractor Dealers, or of Fruit and Potato Traders, and to that extent may have less need of commercial services from their cooperative associations than would be the case elsewhere
- opportunities for the formulation of common policies. Although agricultural cooperatives are, as just mentioned, often members of trade associations covering the non-cooperative as well as the cooperative sector, there are certain questions which arise on which they may find it necessary to develop a separate policy, occasionally in opposition to non-cooperative organizations. This they are able to do through the specialist panels or committees for different commodities which their central bodies have established
- representation on political issues. This function has become increasingly important as the role of government has extended, and actions are taken from day to day which can have an important impact on the trading situation. The cooperatives rely on their central bodies not merely to alert them to im-

peding changes but even more to exercise a degree of prescience in these matters and to take action in sufficient time to ward off possible future dangers. Such 'pressure group' activity can also be helpful to the government, which might not always be aware that the cooperatives had a special case if there were not a central body to present it for them.

The four central associations for England, Scotland, Wales and Northern Ireland all have development responsibilities included among their objects. These have to be exercised with due regard to the interests of their existing member organizations which, as main subscribers to the funds of the central bodies, will naturally expect their own interests to be taken care of. The rules of Agricultural Cooperation and Marketing Services are the most specific on this point, the relevant clause reading as follows:

> To work closely with Government agencies and others in the promotion and development of agricultural cooperation and where possible to promote development through existing channels.

Apart from the consideration of members' interests it is clearly desirable that, wherever possible, existing channels should be used, so that the potential strength of cooperation is not weakened by being divided between a host of separate and, it might be, competing organizations.

The relationship between farmers' cooperative associations and farmers' unions is never an easy one. For the Unions, which have many members doing business in other ways than through cooperatives, cooperation is only one method, among others, for reinforcing the commercial solidarity of farmers. For the Cooperative associations, the interests of their member organizations will not always be identical with those of the general body of farmers, particularly those who have not accepted cooperative objectives and disciplines. The differences of point of view are, however, confined to a fairly small number of matters, there being many more on which farmers' cooperatives and unions have a common outlook and can provide one another with effective support. In England it would appear that a clear recognition of the problem may have gone a long way towards solving it. The decision to establish Agricultural Cooperation and Marketing Services Ltd in 1972 was taken there following the recommendations of a Working Party which had sat to study the future relationship between the former Agricultural Cooperative Association and the National Farmers Union. This Working Party recommended that the NFU should withdraw from its commercial involvements and concentrate on its political role as the representative body for the industry. When the new body was set up it was recognized that

> the NFU is the main political force in the industry and must be built up as the body representing farmers and growers, both individually and 'in cooperation'. The NFU is firmly established as a formidable pressure group and to build up ACMS as a totally separate representational body would both waste resources and reduce the Union's own effectiveness. Whenever possible, therefore, ACMS, in safeguarding its members' interests, acts in close consultation with and often through the National Farmers Union.
>
> Nevertheless, it is clearly recognised that there are and will be occasions when the interests of the NFU and ACMS membership diverge and that in such circumstances it is necessary for the two organizations to put their different points of view separately to Government or elsewhere.
>
> Representation, of course, is not confined to the political sector. There are many instances where technical and commercial issues affecting cooperation need to be presented in conjuction with other organizations sharing a common viewpoint with ACMS.

This passage is taken from an explanatory memorandum about the new organization issued in 1972.

During 1974 a proposal was being considered for the establishment of a national Council of Agricultural Producers, which would have the same membership as BAMDO (see Chapter 2), but wider terms of reference. These would include coordination of the interests and activities of the member organizations – farmers unions, marketing boards and cooperatives – so as to avoid conflict between them and present a united front.

The entry of the United Kingdom into the Common Market meant that the agricultural cooperatives had to create machinery which would enable them to play their part in the General Committee of Agricultural Cooperation of the EEC countries

(COGECA) in order to take advantage of the direct access this body enjoys to the Commission. Such liasion could only be arranged on a national basis and it was therefore decided that the Federation of Agricultural Cooperatives – a body which had existed previously mainly for the coordination of internal policies – should take on the additional responsibility. FAC has therefore become the UK cooperative representative on COGECA, which in turn has the same sort of close liasion with COPA, the body representing the European farmers' unions, as that which now exists between cooperative associations and unions within the UK.

Apart from the four national associations and FAC, not many central organizations are to be found in the United Kingdom, for reasons which have already been discussed. Various federal trading organizations have been mentioned in Chapters 8–9. It will be appropriate to mention here the Agricultural Credit Corporation and the Agricultural Finance Federation. The former is not a cooperative, but is a body formed in 1959 by the farmers' unions of England and Wales, Scotland and Ulster to guarantee the borrowings of individual farmers and growers, so that they might be able to obtain additional Bank credit. The latter is a federal organization jointly owned by the Cooperative Wholesale Society, which owns half the shares, and some thirty agricultural cooperatives, which own the remainder. This Federation has been concerned with the business of hire purchase, machinery leasing and discounting bills for agricultural cooperatives, and it too is empowered to give guarantees. The role of these organizations in connection with the official Credit Guarantee Scheme has already been discussed in Chapter 4.

The last organization to be mentioned under this heading is the Plunkett Foundation for Cooperative Studies, a body set up by Horace Plunkett under a Trust Deed of 1919 in order 'that there should be greater facilities for the systematic study of the principles and methods of agricultural and industrial cooperation, in which lie possibilities of great promise for the future well-being of the rural community and of the nation as a whole. . . .' The Plunkett Foundation's engagement in training and research is partly in overseas countries, but in recent years it has also become an agent for research and for the compilation of basic statistics on behalf of the Central Council and the Federation of Agricultural Cooperatives in the United Kingdom. Through membership of CEA (see Chapter 12) and links with the International Cooperative Alliance, the Foundation maintained what were, for many years, almost the only contacts between the United Kingdom and agricultural cooperative movements overseas.

CENTRAL COUNCIL

It has already been noted (in Chapter 2) that the decision to give more support to agricultural cooperation was part of a wider policy for the Development of Agriculture outlined in the White Paper bearing that title published in August 1965. It was in this document that the proposal to set up a Central Council appeared, though the Council did not actually come into being until two years later, on account of the delay in the legislative programme caused by the 1966 general election. The thinking behind the proposal can be traced from various phrases used in the White Paper. The Central Council would administer grants on behalf of the Agricultural Departments, it would have considerable powers of discretion, it would cover the whole of the United Kingdom, it would be able to work through existing cooperative bodies, it would be responsible for discovering and popularizing all kinds of cooperative activities, it would supply purposeful direction and coordination to what would otherwise be (and had been in the past) no more than a piecemeal approach. The points that the White Paper appeared to be making were that the discretionary powers needing to be attached to the awarding of cooperative grants would make it difficult for these to be administered directly by agricultural departments, as had been done on a small scale in the past, while the absence of any single national cooperative authority meant that existing associations could not be used to do the job, though they could be employed as agents. In fact, had the government at the time decided to make use of existing national bodies it would have had four if not five of them to deal with, mostly made up of organizations of the traditional type, i.e. whose farmer members were not usually bound to them by any form of agreement. Such bodies could hardly be expected to be enthusiastic about the ideas that the government had in mind to encourage.

These were perhaps the dominant reasons which persuaded the government of the necessity to establish a new independent statutory corporation, rather than attempt to use or adapt any existing machinery for its purposes.

The question of farmer member – agricultural cooperative relationship was fundamental to the new thinking. The government stated that no distinction would be made between one type of trading body to be aided and another

provided that the body had a definite constitution of a cooperative character. This constitution must provide in each case that the body concerned possesses, and shall make use of, powers to ensure that members fulfil certain minimum obligations and are loyal to it. Experience has shown that cooperative bodies that do not exercise this form of discipline run into one or other of two dangers. They may collapse through lack of consistent support, or may become little more than wholesaling firms of which farmers happen to be the shareholders.

The existing agricultural cooperatives in the United Kingdom were at first very much taken aback by this statement, not so much because few of them were able to meet the government's novel definition of what constituted a cooperative as by the grounds alleged for introducing it. These they did not accept as being true, and read as being critical of themselves. Their reactions to the government's policy were not at first very warm, therefore, though in the course of time, as the need for a disciplined approach to marketing became more generally recognized, the intrinsic merits of the policy were widely admitted, and the context in which it had been announced was forgotten

In other respects, bearing in mind the well-known difficulties of introducing new organizational machinery into a conservative agricultural setting, the new Council got off to a good start. This was partly due to its composition, since eight out of a total of fourteen members were appointed by Ministers from a list of names submitted by organizations representing the interests of farmers and farmers' cooperative associations, partly because the staff of the Council was selected (as the White Paper had hinted that it would be) from persons already experienced in this field who had the confidence of cooperatives, but most of all perhaps because among the earliest successful applicants for grant aid were 'traditional' cooperatives, which had, after all, found it not too difficult to adapt themselves to the new criteria. In total, by far the greater amount of grant aid recommended by the Central Council under its grants scheme and approved by the government has been issued to cooperatives which were in existence before a Central Council had been set up or even thought of.

The functions of the Central Council were laid down by the Agricultural Act 1967. The first was

> to organize, promote, encourage, develop and coordinate cooperation in agriculture and horticulture.

and the second

> to put themselves in a position to advise Ministers on all matters relating to cooperation in agriculture and horticulture.

This second function of the Council might appear to risk some overlap with that of central cooperative associations in making representations to the government on behalf of their members. In practice this has not been a problem, as the Council invariably consult the associations before offering advice to Ministers on any important subject. Furthermore, the central associations and farmers' unions always have knowledge, through their nominees on the Council, of matters under discussion there.

The first function of the Council really divides itself into two parts, in that there are certain cooperative activities which are, and others which are not, eligible for aid under the government's Agricultural and Horticultural Cooperation Scheme. This scheme was published as a Statutory Instrument (No. 1329) in 1967, superseded by a revised Statutory Instrument (No. 415) in 1971. The important paragraphs in both schemes are those concerned with eligibility of proposals and eligibility of applicants. So far as the former is concerned, proposals may be recommended by the Council for financial aid, provided the grant would not be made for purposes which would assist the recipients in activities connected with the supply of goods; that is to say the Scheme is limited (with minor exceptions) to the encouragement of cooperation in production and marketing. As regards the eligibility of applicants, the Council must be satisfied that the applicant is a cooperative both in the general sense and according to the special criteria of the White Paper, and must examine the extent of the potential benefit to producers, the scale and general merits of the proposal, including its practicability and suitability for the applicant, how the proposal will be carried out, and what are the applicant's resources. In order to meet these conditions the Council has, over the years, worked out detailed guide lines of what proposals can be recommended and will be acceptable to Ministers, with whom the final decision rests, so that applicants may be advised at an early stage whether projects which they are considering would come within the scope of the Scheme.

In order that the Council may be able to use its discretion the Scheme specifies only what shall be the maximum rates of grant. The 1971 Scheme brought cooperative capital grants more closely into line with those obtainable by individuals under other grant schemes, but it still includes special grants towards feasibility studies and the employment of key management staff, which are available to cooperatives alone. They are invaluable forms of aid. Insistence on feasibility studies has meant that the applicants are associated in the preparation and planning of the enterprise and thus far better prepared to undertake it. Grants towards management costs help to ensure that the right kind of person is employed, so that the venture will get off to a good start.

As regards the capital grants for works and facilities, calculations made by the Working Party on Agricultural Cooperative Investment, whose report is referred to in Chapter 4, suggest that these grants have played an important part in the total investment made by marketing cooperatives in recent years. Much of the development in horticulture, in particular, appears to have been due to this stimulus.

The promotion and development functions of the Central Council contained in the 1967 Act cover all agricultural and horticultural cooperation, not just those of their activities which can qualify for government grants. The Council has accordingly had to involve itself in planning over a wide field, and with all the conditions under which cooperation has to operate. As mentioned in Chapters 3 and 4, detailed studies have been made of the general legal and financial problems of cooperatives, as well as special surveys of separate sectors of cooperation, referred to in Chapters 7, 8 and 9. As always, there is a problem of disseminating knowledge among those who are able to benefit from it. The White Paper spoke of promoting and popularizing all kinds of cooperative activities among farmers. This promotion and popularization has been the particular responsibility of the Council's regional staff who, being closely in touch with sources of information in the areas which they serve, are able to intervene at an early (and therefore often decisive) stage, when they hear of a cooperative project among farmers being mooted. In this connection it is interesting to note that, despite all the past efforts that have been made to propagate basic ideas of agricultural cooperation among farmers, these are far from being generally understood, so that regional officers have still to spend as much of their time in explaining the principles as they do in working out the practice of cooperation. The means of promotion used consist of seminars for farm advisers, farmer cooperative staff and farmers, particularly those serving as cooperative directors, meetings, farm walks, agricultural shows, films and publications, which include the quarterly journal *Farming Business* produced by the Central Council in conjunction with the central associations. Regional officers also have supervisory responsibilities, that is of seeing that grants have been correctly spent, and that projects are being developed in line with the plans approved for them. They can give the management services branch of the Council early warning of any special difficulties which a newly formed cooperative has encountered, so that special assistance can if necessary be rendered before it is too late.

As in its function of advising Ministers, so too in that of organizing, promoting, encouraging and developing cooperation, the Central Council runs the risk of overlapping with central cooperative associations. This risk is avoided partly by the processes of consultation mentioned earlier, and partly by the central associations themselves undertaking certain development activities as agents of the Central Council. To some extent, such delegation of its functions by the Council had been the practice ever since it was set up. In 1973 the process was taken a stage further, when the government of the day announced its intention of making additional finance available to the Central Council, a portion of which was to be used in fortifying the marketing development capabilities of the central cooperative associations.

The '1974 Fund', as this new source of finance came to be known, is accordingly divided into two parts. The larger share has been allocated to meeting the cost of the new development posts, or making a contribution to the cost, in the case where a post is on the way to becoming commercially self-supporting. The development officers concerned report to their own central associations, each of which has a separate agency agreement with the Council covering the activity in which it is engaged. The Council's own field staff of regional officers will at the same time become more exclusively concerned with promotion, along the lines described earlier. The other part of the 1974 Fund will be used to establish a specialist marketing unit within the Central Council. To guide this unit, and to provide a better liaison at levels where the development work is taking place, a Marketing Policy Committee has been set up by the Council, and given the further responsibility of advising the Council on its agency arrangements. In this way the Committee will be involved in all matters connected with the 1974 Fund. The majority of the members of this Committee have been nominated by the agricultural industry (initially by BAMDO, but in future by its successor organization), and represent a range of different commodity interests. Their role is therefore distinct from that of the industry's representatives on the Council itself, who have been appointed by Ministers to serve there, on the advice of the cooperative associations and farmers' unions.

At the time of writing, the system outlined in the preceding paragraph was still only partially operative. Clearly, it represents a compromise between different courses of action which might have been followed, and it sets some problems for the future. The cooperative associations, while honouring their agency agreements with the Central Council, have to show the independence expected of them by their members. Also, the general responsibilities for development accepted by them may on occasions be difficult to reconcile with the aims of established cooperatives to pursue their own policies in their own way without interference. The Central Council, too, has been given a delicate task of introducing and implementing a strategy of development through field staff who will not be under its control. But it is only fair to state that any other compromise would have resulted in anomalies, of a different kind perhaps, but no less severe. There is general confidence that the system actually introduced can be made to work satisfactorily at the present time, though it is reasonable to assume that some further adjustments will be called for as the situation develops.

11. Relations with other business sectors

Wherever old farm account books from the early part of the present century have been preserved they usually evoke the same comment from modern readers – how little had to be bought in from outside the farm to support the resources of land and labour, at a time when agricultural productivity was comparatively low, and muscle-power of horse and man was still the main source of energy. In those days farmers could be accurately described as 'primary producers'. How their situation has changed since then can be judged from the present national farm account, as set out in the *Annual Review* for March 1974. The figures, which have been quoted in a footnote to Chapter 4, show that the value of the agricultural input is not far short of half of the ultimate value of the produce. These figures make it clear that farmers now occupy an intermediate position, between the suppliers of agricultural inputs, such as feedingstuffs, fertilizers, fuel, etc., on the one hand, and the wholesalers, processors and distributors of farm produce on the other, the three sets of activities including their own being grouped together into what is sometimes referred to as the Food Industry, although that designation is as often used more narrowly to describe particularly the firms making up the third sector. It should be pointed out that the separation into three sectors or stages is normal but by no means essential; it is possible and it also happens in some cases that the manufacture of inputs, production, and the marketing operation are all performed by a single firm. In other cases there may be more than three stages, and more than three independent operators involved.

This chapter is concerned with the situation of farmers' and growers' cooperatives in relation to their non-cooperative market partners in a Food Industry (in the wider sense) where the functions mentioned in the last paragraph are being carried out by independent operators who, however, have become increasingly aware of their interdependence, and who are actively searching for a way of overcoming the divisive factors in their relationship. In other words, they seek to collaborate, in such a way as will allow each of the parties concerned to receive a fair return on his labour and investment. Combined operations of this kind are different from and perhaps more difficult than the cooperation that has been discussed in earlier chapters, which involved participants who, though still independent of one another, were all engaged in broadly the same type of business activity. The difficulties are first that 'where transactions are settled on the basis of a balance between supply and demand, there is a natural tension between the attitudes of the seller and the buyers', secondly that 'agriculture is, by nature, an industry where supply is unstable . . .; matching a variable supply to a fairly consistent demand clearly poses a problem for the food industry over and above what would be considered the ordinary hazards of trading', thirdly that 'there are inherent difficulties in describing or measuring in scientific terms certain characteristics in natural products, and also there may be considerable variations in such characteristics from season to season', and fourthly that the non-farming part of the food industry is 'admittedly oligopolistic', in contrast with farmers, who are sole traders and whose individual economic power is negligible. The above excerpts are from the report of the Committee of Inquiry on Contract Farming (Cmnd 5499) under the chairmanship of Sir James Barker, which was presented to Parliament in October 1972. The commissioning of this report was an indication of the importance attached by the government to the matter of relations of the component parts within the total food industry and may – if a recommendation made by the committee is accepted – give rise to sample surveys of contract farming to be repeated at regular intervals, along the same lines as the one contained in this report.

The Barker Committee examined the benefits as well as the difficulties of contract farming, from the separate points of view of the producer, the processor or distributor, and the farm supplier. To sum these up, it appeared that the desire for the security offered by long-term or permanent relationships is related to the extent of the investment which each party has to incur. Conditions are ideal for the integration of commercial policies where the various parties concerned have all made a substantial investment, which will only be profitable to them if the facilities they have established are used to full capacity. (It may be recalled that a report on *Cooperation in Pig Production and Marketing*, referred to in Chapter 8, came to precisely the same conclusion.) Where one of them has not had to make such an investment, it is to be expected that his commercial policy will be a good deal more opportunist.

The Committee recognized various types of commercial relationship. The principal divisions are normal buying and selling, contracting, joint ventures, and vertical integration. The involvement of farmers' and growers' cooperatives in each of these will now be considered.

NORMAL BUYING AND SELLING

The Committee made the interesting observation that contracting is appreciably less extensive than in the UK, in countries where cooperation is more firmly entrenched (notably in the Netherlands and Denmark), but it is not clear whether they appreciated the essential difference between a contract by a farmer to his cooperative and a contract by him to a private firm. If a cooperative is able to rely on the loyalty of its members (which may be ensured by a clause in the cooperative constitution, or by means of an agreement separate from the constitution, or simply exist as a matter of fact) the cooperative 'adds value' to the goods or produce it handles on their behalf, in terms of volume if nothing else, and thereby exercises a degree of market power vis-à-vis the supplier or purchaser which may not need to be reinforced by a contract, as would have been necessary in the case of a series of transactions by individual farmers or growers. A practical illustration of this principle is to be found in the condition (referred to in earlier chapters) required for grant-aid, under both UK and EEC schemes, that the producer should be committed to his producers' organization (or group, or cooperative), not that this organization should in turn be committed to a supplier or buyer, though, of course, it may itself decide to become so.

CONTRACT BUYING AND SELLING

This was defined by the Committee as consisting of 'systems for the supply of agricultural or horticultural produce under forward contracts, the essence of such arrangements being a commitment to provide an agricultural commodity of a type, at a time and in the quantity required by a known buyer'. The general picture to emerge from a survey undertaken by the Committee was of 40 per cent of total UK output being disposed of under written contracts or other formal direct selling arrangements, of which, however, 29 per cent consisted of commodities covered by statutory or other centralized marketing arrangements (notably milk, sugar beet and cereals), leaving 11 per cent covered by privately arranged commerical contracts. Important items in this figure were fat pigs (43 per cent contracted), poultry (29 per cent), eggs (23 per cent), fruit (15 per cent) and vegetables (11 per cent). Unfortunately these findings do not distinguish between contracts to a cooperative and contracts to others, mainly processors, though the distinction is really a vital one. The contract situation 'often involves some degree of transfer management responsibility and with it some degree of transfer of the commercial risk'. The Committee's preference was 'for contracting by grouped enterprises rather than by individual producers. This is not to say that the individual farmer – especially one in a large way of business – should be other than entirely free to operate independently if he considers that it is profitable to do so. In general, however, through the workings of economic power, the balance between the benefits and hazards of contracting is likely to be progressively improved the larger the volume of produce covered by the contract. In addition to this, of course, other benefits will accrue through the greater command of resources by group enterprises.'

The survey distinguished between buying contracts, marketing contracts and 'transferred management' contracts, these last being arrangements for part or all of the farmer's managerial function to be transferred to the other party. The Committee concluded that transferred management contracts were less likely to be equitable than simple marketing contracts, for the reason that it was almost impossible for risks to be transferred in equal measure to the transfer of functions. Obviously, cooperatives are less likely to enter into transferred management function contracts than are individual farmers, though cases of their doing so are not unknown. The division into buying and marketing contracts is over-simple. Many contracts are concerned with both aspects. Such two-way contracts are regarded as involving special risks, as will be noted later in this chapter.

JOINT VENTURES

Contracting is an 'arms-length' relationship, which can be broken off when it becomes inconvenient. Joint ventures involve a more permanent relationship, in which the link is constitutional rather than contractual. It is interesting to compare the French and the UK approach. In France a special constitutional form has been created, that of an Association of Collective Agricultural Interest (SICA). This was done in order to overcome a difficulty from which the agricultural cooperatives originally suffered, though it has since been removed, that they were not able to transact any non-member business. As a result, a number of purely farmer-owned businesses found it more convenient to register themselves in this form. Others were set up with a membership consisting partly of farmers (or farmers' cooperatives, etc.,) and partly of non-farmers whose business was of a kind to facilitate the realization of the objects of the SICA. Such non-farmer members would normally be firms engaged in the distribution or processing of agricultural produce. In a SICA, as in a cooperative, there is a limit on the rate of distribution of profits to be made on the share capital. There are also important restrictions on the voting power of the non-farmers in relation to the farmer members. Nevertheless there is still the advantage, to the non-farmer member, that a SICA has the privilege of belonging to a Bank of Agricultural Credit. In the United Kingdom policy has developed more along the lines recommended by the Barker Committee, which were that it was better for grouped enterprises, i.e. cooperatives, rather than individual producers, to be involved in such 'interprofessional relationships', as they are described in France. Thus, when the scope of the grants available through the Central Council was extended in 1973 so as to assist joint ventures, the condition was laid down that a cooperative and a private firm should set up an independent company for the purpose of the joint activity, in which the cooperative's stake would have to be not less than 50 per cent. Up to date, little if any use has been made of this facility, though a good deal of interest has been shown in it.

VERTICAL INTEGRATION

This has been defined as the combination of two or more successive stages in the production/processing/distribution chain under the ownership of a single corporate, cooperative or private, venture. In practice control can be virtually complete without ownership and, as mentioned earlier, firms supplying inputs may also be involved. Agricultural spokesmen, reflecting to some extent the anxieties of the rank and file of farmers, have been very sensitive to the risks of creeping integration, resulting perhaps in agricultural producers losing effective control over their own businesses. As far back as 1964 a National Farmers Union study group reported on the matter. Its main conclusion was that the real solution lay in integration under the control of the producer. But already, even by the time the report was published, such a recommendation had ceased to be practicable. Heavy investments had been made by private and public companies, in other parts of the food industry, in facilities for manufacturing farm supplies and processing farm produce while, so far as farmers were concerned, the absence of direct government aid and other measures had, in the opinion of the Barker Committee, 'tended to condition farmers to the belief that their investment responsibility ends at the farm gate'. It is true that the Marketing Boards, some companies under farmer control, and the Cooperatives, have important investments outside the farm gate, but these represent only a small proportion of the total. If therefore agricultural producers are to achieve a satisfactory relationship with other parts of the food industry, this may have to be developed by techniques of contracting or joint ventures rather than by vertical integration under the control of the producer or, more likely, of some business outside farming.

PROBLEMS OF INTERPROFESSIONAL RELATIONS

The report of the committee of inquiry on contract farming was the first serious attempt to consider these problems and it was necessarily incomplete. In the first place it did not establish trends (though it did suggest that these needed to be investigated) and, secondly, it said very little on the subject of joint ventures. There is no doubt that agricultural economics departments of universities could, as the report suggests, play an important part in monitoring progress and suggesting new ideas in this important but hitherto somewhat neglected field.

Other solutions to the problem depend to an important extent on the attitude of the government to the general question of maintaining an equitable relationship between agriculture and other sectors of the food industry. There is of course a political commitment to this contained in Article 39 of the Treaty of Rome, as well as in the Agriculture Act 1947, and it is generally believed that the subject of interprofessional relations in agriculture is one on which the Commission of the EEC has an intention to propose legislation. In view of this, it will be useful to consider what are the principal problem areas.

First, there is a widely held view that farmers should be protected from entering into contractual or constitutional relationships with other sectors of the food industry of a type which could prove disadvantageous to them. In both French and British legislation, as we have seen, grant aid has only been made available for joint ventures where a 'safe' proportion of the voting power is held by or on behalf of farmers. Where contracts are concerned, French legislation goes further, and enables farmers to repudiate arrangement where the person supplying the input is also responsible for organizing the sale of the output, unless the nature, price and quality of the supplies have been specified in the contract. In the United Kingdom farmers and their organizations are not protected in this way but they receive help from their own representative bodies in the drawing up or scrutiny of contracts, with government contributing towards the cost of this service.

Secondly, agricultural cooperatives in a number of countries fall foul of restrictive trade practice legislation which, if strictly applied, would make it virtually impossible for them to enter into external trading relationships. Although in the United Kingdom the relief given may have been less complete than the cooperatives would like, their special problems have been recognized and dealt with under an Act of 1962, as mentioned in Chapter 3.

Thirdly, one of the chief obstacles to the development of forward contracting or joint ventures is the lack of a satisfactory method of determining forward prices. The subject is too technical for examination here, but briefly there are two main methods, formula pricing and profit sharing. The French law on interprofessional agreements indicates that prices should be fixed by reference to costs of production, which is a one-sided form of formula pricing; superficially this is attractive, but not likely to prove so in the long term. Profit-sharing is the more satisfactory alternative, but is tremendously dependent on a high degree of mutual confidence.

Fourthly, the Barker Committee concluded, and there has been a good deal of support for their view, that the commercial knowledge available to farmers' organizations about the intentions of processors and distributors, and the information possessed by the latter about the potentialities of the former, was generally incomplete. They recommended therefore that one of the functions of an Agriculture and Food Development Authority which they proposed should be established would be to bridge this communications gap. (The Committee thought that this might also help to deal with situations of over-supply, but that is a more doubtful proposition.) Although the government did not accept the idea of a food industry authority, this particular one of its suggested functions will be a responsibility of the marketing unit of the Central Council referred to in Chapter 10.

Fifthly, the Barker Committee looked to their proposed Authority to provide a new resource base, both financial and management, for agricultural production and marketing. Their analysis of the need was as follows. 'The individual farmers who are members of a group probably possess enough wealth to cover the risk element in group activities, but it is rare for the group as such to own the financial resources to meet group needs. The tradition of supporting cooperative ventures as an integral part of the investment in a farming business is significantly less strong in the UK than elsewhere in Europe and this – coupled with other factors mentioned earlier – means that the first tier bodies themselves suffer from an essential lack of creditworthiness. This, in turn, becomes a serious impediment to progress at the second tier when major capital projects are contemplated either as sole or joint ventures.' The response of government to this statement of the Committee's view was to invite the Central Council to set up a working party to study the adequacy of capital investment in agricultural cooperatives. Reference to the examination made by this working party has already been made in Chapter 4.

NOTE: An additional study of the problem of vertical integration in agriculture is contained in a survey made by M. W. Butterwick in 1969 and published by the Central Council. An authoritative statement on the application of the EEC laws of competition to producers' organizations is contained in Section II of *Les coopératives agricoles dans le marché commun* by J. G. de Villeneuve and others, a translation of which may be obtained from the Central Council. The special situation relating to interprofessional bodies and joint ventures in France is covered by a report made for the Central Council in 1974 and obtainable from them. Finally, for a useful account of the problems of price fixing in inter-professional relationships reference may be made to an article by Professor Goldberg entitled 'Profitable partnerships: industry and farmer co-ops' in the *Harvard Business Review* of March–April 1972.

12. The EEC and the cooperative future

Some of the distinctive features of continental cooperatives have already been mentioned. Their membership commonly consists of farmers who are more closely committed to their cooperative, both constitutionally and commercially, than is the case in the United Kingdom. These cooperatives account for a large proportion of the total agricultural business in their countries, and have substantial investments in the manufacture of farm inputs and the processing of farm products. They depend for a major part of their finance on specialist credit institutions which, though no longer concerned exclusively with agriculture, still treat it as a privileged industry. In most of these European countries there has been a fair amount of State intervention in favour of agricultural cooperatives, with a correspondingly wide range of national legislation. It was never the intention that this should be superseded, when the European Economic Community was set up. Rather the existence of cooperatives is taken for granted by the Treaty, which mentions them at only one point, in connection with the right of establishment (Art. 58). Nor was it considered necessary to refer to cooperatives when defining the scope and objects of the common agricultural policy (Art. 38). Indeed it would have been inappropriate to do so, seeing that there are numerous important agricultural cooperative activities – such as the supply of farm requisites, credit and insurance – which that policy does not cover.

COOPERATION IN THE COMMON AGRICULTURAL POLICY

It was however to be expected that in elaborating the common policy, as the Commission was required to do (Art. 43), there would be some definition of the part that cooperatives were expected to play in it. This duly appeared, in mid 1960, in a memorandum (Doc. VI/COM(60)105 of 30.6.1960) which has been quoted in Chapter 1.

Again in 1962, in a further memorandum (Doc. 8067.1-XI-1962–5) setting out its programme of action, the Commission expressed the view that agricultural cooperatives should be encouraged.

The common agricultural policy in respect of prices was completed in 1969. In respect of structure it is not complete even now. The earliest Community Regulation on this subject, brought into force in 1964 (Regulation EEC/17/64), was couched in general terms, which gave little guidance as to the mechanisms to be adopted in later more specific legislation. When the later Regulations, or draft regulations, began to appear, it seemed that there must have been some change in the Commission's earlier thoughts on the subject. Instead of specifying 'cooperatives', the new texts dealt with 'producers' organizations' (Regulation 1035/72) as the bodies to be encouraged to undertake the marketing of horticultural produce (and intervention), and 'producers' groups' to perform a similar task in agriculture. The latter Regulation still remains a draft, however, as it has not yet been approved by the Council of Ministers. Only a single crop, hops, has been covered by a producers' group regulation (Regulation EEC 1696/71).

These 'organizations' and 'groups' have two principal characteristics. They have to be composed of producers of crops which the organization or group is to handle, and the whole of that production must be sold through that body. What is missing is any cooperative provision that the members of the organization or group will receive benefits in proportion to their throughput, or that control over it will not be exercisable by a minority. The reaction of existing cooperatives in the Six to the new structural policy was predictable. They felt that it ignored the realities of the situation, by encouraging the formation of new producer bodies of doubtful worth instead of building on existing foundations. In practice, however, this does not appear to have happened yet to any appreciable extent.

The Commission's ideas in respect of the structure of agricultural marketing are still tentative, and have not as yet appeared in any draft Regulation or Directive. They appear likely to make use of the concepts of inter-professionalism elaborated in France (see Chapter 11) and, in particular, to assess the merits of the relationships worked out between producers and other sectors of the food industry on the basis of the contracts entered into and what they contain. Discussions on these matters were in progress at the time of writing but, given the virtual standstill in the development of the common agricultural policy, there can be no expectation that they will be brought to any speedy conclusion.

Community policy has shown little interest in the potentialities for cooperation in production. Community legislation (Directive 72/159) dealing with the modernization of farms made a gesture in favour of encouraging groups whose aim is mutual aid between farms, a more rational pooling of agricultural equipment, or joint farming, but the incentives offered are fairly modest. In another directive, still in draft, dealing with the special problems of difficult areas, the cooperative provisions are somewhat stronger. In view of the powerful sentiments evoked in all the original member States by the concept of the independent family farm, it is not surprising that the main emphasis in EEC legislation should be on maintaining their independence.

COGECA AND CEA

While the established cooperatives have had some cause for disappointment concerning the trends of structural reform in the EEC up to date, and also concerning the relatively small proportion of the European Agricultural Guidance and Guarantee Fund allocated to the reform of structures (the 1973 budget estimated an expenditure of £146 million on guidance purposes out of a total of £2083 million) they have at least some assurance that they will be consulted about the evolution of the common agricultural policy in so far as it may affect them. In 1960, when making the pronouncement mentioned earlier in this chapter, the

Commission set up a series of consultative committees, through which to seek the advice of the various professional organizations. In preparation for this development, the leading national agricultural cooperative organizations had already formed an apex organization, or general committee of agricultural cooperatives of the member States of the EEC (COGECA), which was allotted a certain number of seats on each of the consultative committees. As mentioned in Chapter 10, the agricultural cooperatives in the United Kingdom have taken up membership of COGECA and, through it, are in direct contact with and can bring their influence to bear on the Commission in Brussels.

COGECA has a number of committees, on which the experts in various agricultural sectors from cooperatives of the different member States can meet to discuss with one another the impact of Community Regulations or Directives in their particular field and try to work out a common policy. It is also to some extent an instrument of longer-term planning. For instance it has devoted a considerable amount of time to the drafting of a Statute for European Cooperative Societies, with the idea that if this could be introduced in EEC law alongside a Statute for European Companies, it would provide a base for the international cooperative societies which will be needed, following a closer integration of trade in agricultural products between member States.

There should also be mentioned in the same connection a more widely based, loosely knit international body, known as the European Confederation of Agriculture (CEA), with headquarters in Switzerland, of which a number of the central cooperative associations in the United Kingdom are members. For many years the Commission (i.e. committee) of this Confederation concerned with agricultural cooperation, mutuality and credit has provided a forum through which the representatives of agricultural cooperatives in Western Europe have exchanged ideas with the object of formulating united policies and common action. The development of an international cooperative trading agency for cereals, mentioned in Chapter 8, was one of the results of the valuable opportunities for meeting provided by membership of this Confederation.

APPLICATION OF EEC LEGISLATION IN THE UK

The legislative processes of the European Communities tend to be laborious and slow, particularly where structural aspects of the common agricultural policy are concerned. Likewise, once a policy has been agreed upon and is incorporated in Community law, it becomes difficult to alter. Nor is the policy capable of much adjustment, even where the legislative instrument has taken the form of a directive, in order to meet the special circumstances of individual member States.

Since these are the conditions which will apply in the future, if past experience can be taken as a guide, it becomes imperative that all the parties affected by them, above all any agricultural cooperatives concerned with production or marketing, should consider very carefully what changes they may need to make so as to satisfy the requirements of this legislation. (It has to be assumed that the present draft of the legislation proposed for producers' groups will eventually be passed, if not in the same form, at least without any essential change concerning the nature of the bodies that are to be given recognition.) As already mentioned, these requirements are that the body of producers to be recognized must consist only of persons actually engaged in producing the crop to be marketed, and that the entirety of their crop must be committed by them for marketing through it. These are sensible provisions in themselves, but there will now be additional reasons for production or marketing cooperatives to adopt them wherever it is possible for them to do so. The first is that only cooperatives so organized will be eligible for the financial incentives to be offered under this legislation. Secondly, it is only bodies of this sort which can claim automatic exemption, as long as they pursue the objects for which they were formed, from the rules of competition (the EEC equivalent of British restrictive trade practice legislation) which might otherwise be applied to them.

In brief, membership of the EEC does not result in any new direction of policy where agricultural cooperatives are concerned, as the EEC rules are very like the disciplines that have to be accepted by cooperatives seeking to establish an acceptable status under the Agricultural and Horticultural Cooperation Scheme, and this has been in force in the United Kingdom since 1967 (see

Chapter 4). What has happened is rather that a line of policy for cooperative development which has been gaining ground for some time can now be regarded as having become more or less stabilized.

LINES OF FUTURE COOPERATIVE DEVELOPMENT

At this point it may be useful to look ahead, and try to envisage the sort of characteristics which should be encouraged in the modern agricultural cooperative wishing to adopt a structure that will be appropriate for the conditions obtaining during the next decade. An indication of the changes that are likely to be needed has been given elsewhere, so it will only be necessary to summarize them. Briefly, they concern the control (by votes), the trading commitment and the financial responsibilities of the members.

(i) A weakness to be found in many of the older cooperatives, and now well recognized by them, is that too many of the members, owning a significant proportion of the total share capital, are persons who for age or other reasons have given up doing business with the cooperative. Whether this situation arose through miscalculation of the resources needed to be put aside for the repayment of shares held by them, or through simply ignoring the problem in its earlier stages, is a matter of past history. What is important now is that cooperatives which cannot foresee an early opportunity to repay these inactive members should encourage them to take up a new status of preferred share holders or loan holders, with a higher rate of interest on their investment in the cooperative, in return for a surrender of their votes. It may be argued that, in practice, inactive members do not exercise their votes. But the risk that they may do so is one that the cooperatives should not allow themselves to face, seeing that it is fundamental to their policies that the control of the organization should rest with those who use it. Cooperatives with a large number of inactive 'financiers' among their membership cannot be oblivious of the fact that, to such persons, the return on their capital is all that matters. They will be bound to shape their policies accordingly, will behave increasingly like ordinary non-cooperative companies, and will understandably come to be regarded as nothing different from such companies by their members.

(ii) Agricultural cooperatives in the United Kingdom have always been founded on the basis of limited liability; in practice this has meant very limited liability, for the average shareholding of the average member is small. Consequently, members have never had any fear of failure by their cooperatives causing grave embarrassment to themselves, and have tended to treat them as ordinary trading concerns, for the success or failure of which they have only a minor responsibility. Now it has become evident that, as part of a much more closely integrated food industry, cooperatives must be able to enter into engagements with other concerns in the full certainty that their members will enable these engagements to be carried out. In place of the unlimited liability for the cooperative which exists over much of the continent, the members of a UK cooperative must be prepared to undertake a contractual liability, which will give it the same or an even better assurance.

(iii) In many agricultural cooperatives, even where members' finance is adequate, it is inequitably divided among them. The fairest division is in accordance with trade done, or capable of being done, but admittedly there are difficulties in applying this principle. One is that different individuals have different financial circumstances, and some of them (particularly the older and longer established) may be able to meet the cooperative's demand for capital more easily than others. Another is that the capital investment requirement may vary considerably as between one item of trade and another. Nevertheless, the principle deserves to to be followed as widely as possible, not only because it is obviously fair to all concerned, but also because it then becomes unimportant whether the return made to members on the basis of the year's trading is calculated on investment or on the amount of trade. At present a cooperative has to decide how much to allocate to dividend on shareholding, within the limits laid down by the Registrar, and how much to bonus on trade. Under the system proposed there would cease to be any essential conflict between these two alternatives.

These three features to be incorporated in agricultural cooperative structures of the future are to be regarded as supplementary to the basic cooperative principles which were listed and discussed in Chapter 3. They are to be seen as the adjustments necessary to enable the cooperatives to meet a new situation, involving such elements as highly organized markets, continuing inflation, and the need to make massive investments, none of which were present when the philosophies of cooperation were originally evolved.

Reforms along these lines are within the competence of even the most traditionally organized cooperatives, as recent events have shown. Their effect would be greatly to strengthen the commercial and social force of the body which adopted them, and to bring up to date a system which, while it worked very well in the nineteenth century, and even in the first half of the twentieth, is less than adequate to the economic conditions of today. Besides the power to carry out these reforms, which already exists, the will must be found to introduce them. Cooperatives which have the necessary courage and determination to make such changes will not, it may be prophesied, lack support from their trading partners, their external financiers and the government, in overcoming any further obstacles that may stand in the way of their successful development.

Index